SpringerBriefs in Philosophy

SpringerBriefs present concise summaries of cutting-edge research and practical applications across a wide spectrum of fields. Featuring compact volumes of 50 to 125 pages, the series covers a range of content from professional to academic. Typical topics might include:

- A timely report of state-of-the art analytical techniques
- A bridge between new research results, as published in journal articles, and a contextual literature review
- A snapshot of a hot or emerging topic
- An in-depth case study or clinical example
- A presentation of core concepts that students must understand in order to make independent contributions

SpringerBriefs in Philosophy cover a broad range of philosophical fields including: Philosophy of Science, Logic, Non-Western Thinking and Western Philosophy. We also consider biographies, full or partial, of key thinkers and pioneers.

SpringerBriefs are characterized by fast, global electronic dissemination, standard publishing contracts, standardized manuscript preparation and formatting guidelines, and expedited production schedules. Both solicited and unsolicited manuscripts are considered for publication in the SpringerBriefs in Philosophy series. Potential authors are warmly invited to complete and submit the Briefs Author Proposal form. All projects will be submitted to editorial review by external advisors.

SpringerBriefs are characterized by expedited production schedules with the aim for publication 8 to 12 weeks after acceptance and fast, global electronic dissemination through our online platform SpringerLink. The standard concise author contracts guarantee that

- an individual ISBN is assigned to each manuscript
- each manuscript is copyrighted in the name of the author
- the author retains the right to post the pre-publication version on his/her website or that of his/her institution.

More information about this series at http://www.springer.com/series/10082

Pierre Uzan

Mind-Body Entanglement

Theory and Therapies

Springer

Pierre Uzan
Fellow at the Catholic University of Paris
Associate researcher at Laboratory of Human and Artificial Cognition (CHArt),
Professor of Physical Science at the French Student Health Foundation (FSEF)
Paris, France

ISSN 2211-4548 ISSN 2211-4556 (electronic)
SpringerBriefs in Philosophy
ISBN 978-3-030-90172-1 ISBN 978-3-030-90173-8 (eBook)
https://doi.org/10.1007/978-3-030-90173-8

This Springer imprint is published by the registered company Springer Nature Switzerland AG
The registered company address is: Gewerbestrasse 11, 6330 Cham, Switzerland

Introduction

We experience in everyday life the intertwinement between our emotional experience and the functioning of our body. For example, each of us can observe that "negative" emotions like fear, anger or anxiety are accompanied by muscle contractions, palpitation of the heart, increasing of blood pressure or digestive disorders. The symmetric phenomenon can also be observed since "positive" emotions like joy, pleasure or enthusiasm generally correlate with healing mechanisms, involving the activation of immune defences via the stimulation of the parasympathetic nervous system which works for maintaining the body at rest. Brain imaging has shown precise correspondences between mental processes and the activation of specific brain areas. For example, it has been shown that visual perception correlates with the activation of some areas of the brain that compute the different aspects of the visual information like the colour, the movement or the spatial orientation (Koch 2004). Even more, experiences of binocular rivalry[1] have shown that "high level" areas of the ventral stream are more specifically involved in conscious visual perception (Leopold and Logothetis 1996).

The existence of strong correlations between mental and bodily processes has been established for a long time, from the ancient Greek physicians to the advent of brain imagery and the systematic studies of psycho-neuro-immuno-endocrinology. This science has shown on experimental grounds that the psychic life is strongly correlated with the whole bodily functioning, which involves physiological, metabolic, endocrine and immune processes. Moreover, this intertwinement is revealed by the existence of psychosomatic diseases whose systematic study began with Franz Alexander (1950), co-founder with Flanders Dunbar of the Chicago School of Psychoanalysis. Alexander pursued extensive research on the relationship between emotional disorders and a number of common somatic diseases such as gastrointestinal disorders, diabetes, skin problems, respiratory problems and

[1] In an experiment of binocular rivalry, two different images are presented simultaneously to each eye of the subject whose perception thus alternates from one image to the other while the physical stimulus remains unchanged. Observing the brain activity associated to the perceptual dominance of an image and the perceptual alternations allows studying visual awareness (Koch 2006).

cardiovascular disorders research that has been successfully extended, as discussed in Chap. 3. However, although strong correlations between mental life and bodily processes have been observed for a long time and are now described with accuracy, the question of their nature remains open (Chalmers 2007; Kim 2005; Searle 1997; Uzan 2013). For example, as intuitively suggested, could these correlations be "causal" connections? Assuming that the concept of causation should not be merely reduced to that of correlation, how could we understand the specificity of such a causal connection between two heterogeneous domains? Or, according to a direction of thought upheld by most contemporary biologists, are these correlations the sign that our thoughts and emotions would be nothing more than neural processes (Crick 1994; Dennett 1991) or the informational processes they give rise to (Tononi 2007; Dehaene and Changeux 2004)? These questions and many others about the nature of the connection between the mind and the body constitute what is called the mind-body problem.

This book shows that this problem can find decisive clarifications providing that the specific correlations taking place between a person's mental and bodily features are interpreted as *relations of generalized entanglement*. Such a model of mind-body entanglement whose main ideas are synthetized in Chap. 3 is built step by step: it results from the philosophical analysis of the mind-body problem presented in Chap. 1, and it is developed within neutral monism philosophy (presented in Chap. 2), it utilizes for its precise formulation the mathematical tools of a generalized version of quantum theory presented in Chap. 4. This generalized version of quantum theory is qualified of "generalized" because it applies, beyond the strict physical realm, to any part of mental or physical reality, which define "generalized systems", and to any of their measurable properties, which define "generalized observables". However, it is essential to underline that this generalized version of quantum theory keeps its *full formalism,* meaning that, as is the case for quantum observables, somatic and mental observables are mathematically represented as non-commutative Hermitian operators of a C* algebra while psychosomatic states are represented as vectors of an appropriate Hilbert space.

This book then explores in Chap. 5 the fundamental (and unprecedented) consequences of this new approach to the mind-body problem in medical science, for understanding and healing mental and somatic diseases. Chap. 5 proposes an explanation of existing alternative approaches to healing and the development of new ones within the mind-body entanglement model. The reader has to notice that these medical investigations rely *in an essential manner* on the presented model of mind-body entanglement, which means that the mind-body entanglement model, which first regards the long-standing mind-body problem, and the proposed medical investigations are closely interdependent, *they cannot be regarded as separated parts of this book.*

Asserting that a person's specific psychophysical correlations are relations of generalized entanglement means that their psychosomatic state cannot be described as the mere juxtaposition of a mental state and a somatic state because the information about their correlations is *an essential and irreducible element of this description.* The concept of mind-body entanglement excludes the idea that the individual is the mere union of two independent parts, mental and somatic, endowed with intrinsic properties, and which would "interact". It is the expression of the organizational closure of the individual since it is the way in which someone's mental and somatic features correlate that determines their psychosomatic organization.

This book first reviews the traditional and contemporary solutions that have been proposed, showing their weaknesses. In particular, Chap. 1 shows that the concept of causation is inappropriate for interpreting the mind-body connection. This point is established on the basis of logico-philosophical arguments taking into account the most relevant contemporary analysis of philosophy of mind, such as those of Chalmers, Nagel, Kim and Baumgartner, among others. As will be explained, a consistent interpretation of the mind-body connection requires both keeping a *phenomenal distinction* between the mental and the physical, in order to avoid the problem of consciousness inherent to reductive materialism, and *rejecting its causal interpretation*. This conclusion echoes with Spinoza's neutral monism philosophy which thus constitutes the general framework wherein the mind-body problem can be coherently and fruitfully tackled. Chapter 2 presents Spinoza's concept of psychophysical parallelism and its more recent reformulations, it also shows some of their limits.

In order to deal more accurately with the specificity of the mind-body connection, Chap. 3 will introduce the notion of the psychosomatic unity of the individual. After briefly presenting the development of this notion, Chap. 3 will highlight its characteristics by referring, in particular, to theoretical and experimental studies on emotions. For, as already emphasized by Descartes and confirmed by contemporary neuroscience (Damasio 1999; Lestienne 2008), emotions sit at the mind-body interface and express themselves through both bodily changes and subjective experience. A first property of psychophysical correlations is that they must give rise to the *inseparability* of subjective life and bodily processes. As shown by a study of emotions, by traditional medicines and as confirmed by psychosomatic, the mental and the somatic features of a person are strongly interdependent—and they are so interdependent that caring one of them can heal the other. Moreover, these correlations are *specific* to each person. That is, they associate the processes taking place in *their body* to elements of meaning drawn from *their personal history* within a psychological and social context. In addition to these properties of inseparability and specificity that have been emphasized by psychosomaticians (Trombini and Baldoni 2005), the psychophysical correlations satisfy two other fundamental properties that have not been considered so far or whose importance has not yet been well understood: (1) as shown by the logico-philosophical analysis of Chap. 1 and confirmed by the study of emotions in Chap. 3, they are *a-causal*, which means that they cannot be interpreted as causal relations; (2) they are *non-local*, which means that, as shown by recent studies on the complementarity of emotional observables and also of somatic observables (Uzan 2016), we cannot assign them a precise value before measuring them. Since these last two properties of the mind-body connection are those of a generalized entanglement relation, we can assert that the psychosomatic unity of the individual *requires* the mind-body connection to be a relation of generalized entanglement. In other words, *mind-body entanglement is the condition of possibility of psychosomatic unity*. This approach to the mind-body problem can thus be described as follows: instead of trying to connect as best we can two series of events, mental and somatic, respectively, *we start from the notion of the psychosomatic unity of the individual* and show that what makes this specific organization possible is the generalized entanglement relation that links someone's psychic and somatic aspects and whose properties meet the requirements mentioned above.

As briefly mentioned above, the concept of mind-body entanglement, which can be tested experimentally (see Sect. 4.4), finds a precise and convenient representation within a general systems theory obtained by generalizing quantum theory. This generalized version of quantum theory, whose basic ideas and formalism was developed by Hartmann Römer and initially presented in the paper by Atmanspacher et al. (2002), keeps its mathematical structure while relaxing all a priori references to the physical world. The notion of "system" is defined as any part of reality which is an object of attention and investigation and that of "observable" as any meaningful property which can be measured or observed in a reproducible way (Atmanspacher et al. 2002; Uzan 2014a, b). A precise representation of the closely related concepts of generalized entanglement and generalized complementarity can then be provided: observables are mathematically represented as Hermitian non-commutative operators of a C* algebra and the psychosomatic state of a person is mathematical represented by a vector of an appropriate Hilbert space built from this algebra of operators. The concept of generalized entanglement is obtained by generalizing the key quantum concept of quantum entanglement, initially proposed by Schrödinger for denoting the non-classical, distance-independent correlations between the parts of a composed quantum system that have interacted in their common past (Schrödinger 1935). It is mathematically expressed by the *non-factorability* of the state-vector of the global system, which means that, for a system composed of two parts, its global state cannot be written as the tensor-product of a state of one of the sub-systems and a state of the other. It must be written as a non-factorable weighted linear combination of product-states defined by the tensor-product of the respective basis of the vector spaces associated with these two sub-systems. Entanglement is the central concept of quantum mechanics that explains non-locality phenomena (Einstein et al. 1935; Clauser et al. 1969; Aspect et al. 1982; Tittel et al. 1998). The concept of generalized entanglement is defined in the same way: the state of a generalized system made up of two parts is entangled if it is not factorable. Generalizing the concept of entanglement to any composite system, whatever its nature, allows a more complete and reliable description of holistic phenomena that cannot be analysed in terms of local causation, like all mind-related phenomena. Concerning our purpose, the psychosomatic state of a person will be represented similarly, by a non-factorable weighted linear combination of product-states defined from couples of specific values of mental and somatic observables.[2]

[2] As explained in Sect. 4.1, these couples of values determine the basis vectors of the tensor-product of the mental and the somatic representative spaces.

This systemic interpretation of the mind-body connection in terms of generalized entanglement has precise applications in the area of health and medical science. As will be explained in Chap. 5, if the biomedical model of health focuses exclusively on the body dysfunctions of the patient, the holistic quantum-like approach presented here describes illness as an imbalance in their psychosomatic state, extending thus Hippocrates' line of thought. Healing a person's illness will thus consist in rebalancing their psychosomatic state, which can be done by changing the relative weights of the specific psychophysical correlations encoded in this state. Following this idea, Sect. 5.2 will present two approaches to healing. A first method directly refers to the mathematical representation of the psychosomatic state of a person: as a consequence of mind-body entanglement, an internal disease classified as purely "somatic" has a mental counterpart to which it is strongly correlated—and reciprocally. This view suggests that treating one of them could have immediate healing "effects" on the other, thus rebalancing the patient's global psychosomatic state by increasing the probability of product-states[3] associated to the relevant healing processes. This direct way of healing can be illustrated by the placebo phenomenon that shows that positive emotions induced by a person we trust are capable of triggering healing bodily processes. In the same vein, we can cite biofeedback, mental imagery and meditation for which somatic "effects" are obtained when the subject is in an appropriate mental state. All these self-healing techniques that are presented and modelled in Sect. 5.3 find an immediate justification within our systemic approach grounded on the concept of mind-body entanglement. Their efficiency requires the patient to reach a mental state whose emotional tonality correlates with the somatic observables responsible for the triggering of the healing processes. Such self-healing techniques can thus be taken as genuine therapeutic resources.

However, this direct mode of healing does not always work in the case of a person whose past emotional traumas prevent them from reaching such an appropriate mental state. In this case, healing requires *restoring or reorganizing the patient's body memory by acting on their body* at the point where these repressed emotional traumas have been imprinted—a subject mentioned in Sect. 3.1 by referring to Reich's vegetotherapy (Reich 1972), Janov's primal therapy (Janov 2000) and Fiammetti's somato-emotional approach (Fiammetti 2004). This second way of healing finds here *a theoretical basis* within our systemic, quantum-like approach to psychosomatic unity. For, as an immediate consequence of the entanglement of a person's psychosomatic state, their body keeps track of the succession of their emotional experiences, which form their subjective life. This bodily inscription results from the successive encoding of psychophysical correlations established during the individual's preparation as a biological, psychological and social subject and constitutes their psychosomatic signature. This theoretical justification of the existence of a body memory provided by our systemic approach is corroborated by Damasio's theory of "somatic markers" (Damasio 1994) and is in accordance with Kandel's

[3] These probabilities are calculated by taking the square module of their components in the psychosomatic state.

theoretical and experimental results according to which the memorization process gives rise to associated changes in the neurons of the brain (Kandel 2006). Along this line of thought, we will also mention that this theoretical justification of the existence of a body memory can be extended to that of a *transgenerational* body memory whose evidence can be found, for example, in Ancelin-Schutzenberger's book (1998).

Like somatic disorders, mental disorders can also be much better understood and healed along with our systemic quantum-like approach, as will be explained in Sect. 5.4. This section begins by mentioning some general consequences of this approach regarding long-standing, fundamental questions about mental disorders. Extending the logico-philosophical analysis of Chap. 1 and referring to recent experimental studies testing the alleged cerebral cause of schizophrenia, it first shows that no somatic causation can be assigned to mental disorder. It also questions the generally accepted sharp distinction between normality and pathology as well as that between mental disorders on which is based their classification in the *Diagnostic and Statistical Manual of Mental Disorders* (DSM). It then presents a model of bipolar disorder based on a two-level representation of the psychosomatic state of a bipolar patient. In contrast to the current knowledge of this psychiatric disorder (Bourgeois et al. 2014), this model can explain the alternation between depressive and manic states, their sudden change and the episodes of acute crisis, and it can even predict the moment of shifting to a new phase of the disease. Moreover, this model suggests that other non-drug healing methods, based on the restoration and the synchronization of biological rhythms, can be successfully developed.

This book will end by proposing, in Sect. 5.5, an explanation of the phenomenon of *distant healing* in terms of (generalized) entanglement swapping. This phenomenon will first be presented by referring to some important experimental studies reported in works by Benor (2001) and Dossey (1993, 2008). Distance healing, which can be seen as a parapsychological phenomenon (Radin 2006: 185–191), is actually a modality of *spiritual healing* for which the healer and the person in need of healing can be spatially or temporally distant from each other. This phenomenon appears as a consequence of the patient's mind-body entanglement and of their emotional entanglement with the healer. Distant healing can be understood as a phenomenon of entanglement swapping where the entanglement of the emotional and somatic observables of the patient is transferred into the entanglement of their somatic observables and the healer's emotional observables *once their respective emotional observables have been entangled*. The success of this healing process thus relies on the capacity of the healer to reach an appropriate emotional state of empathy and compassion for the sick. The way by which this emotional entanglement between two persons, namely the healer and patient, can be realized has already given rise to successful experimental studies and to significant conceptual advances. However, such an emotional entanglement still retains some mystery that future research will need to unravel.

References

- Alexander F. *Psychosomatic medicine: Its principles and applications.* New York: Norton, 1950.
- Aspect A., Dalibard J. and Roger G. Experimental Test of Bell's Inequalities Using Time-Varying Analyzers, *Phys. Rev. Lett.*, 49 (25): 1804–7, 1982.
- Atmanspacher H., Römer H., and Walach H. Weak quantum theory: Complementarity and entanglement in physics and beyond. *Foundation of Physics* 32, 379–406, 2002.
- Benor D.J. *Spiritual Healing, Scientific Validation of a Healing Revolution.* Southefield, Michigan: Vision Publications, 2001.
- Bourgeois M.L., Gay C., Chantal H. and Masson M. *Les troubles bipolaires.* Coll Psychiatrie. Paris: Lavoisier, 2014.
- Chalmers, D. The Hard Problem of Consciousness, in M. Velmans and S. Schneider *The Blackwell Companion to Consciousness,* Blackwell Publishing, pp. 225-235, 2007.
- Clauser J. F., Horne M. A., Shimony A. and Holt R. A. Proposed Experiment to Test Local Hidden-Variable Theories. *Phys. Rev. Lett.* (23): 880-890, 1969.
- Damasio A. *The feeling of what happens: Body et Emotions in the making of Consciousness.* New York: Harcourt Brace, 1999. French trans. *Le sentiment même de soi. Corps, émotions, conscience.* Paris: Odile Jacob, 1999.
- Damasio A. *Descartes' Error: Emotion, Reason and the Human Brain.* New York, Avon Books, 1994.
- Dehaene S. and Changeux J. P. Neural mechanisms for access to consciousness, *in* Michael Gazzaniga et al., *The Cognitive Neuroscience III.* Cambridge: MIT Press, 2004.
- Dennett, D. *Consciousness Explained.* Boston: Little, Brown, 1991.
- Dossey L. Healing Research: What We Know and Don't Know. Explore Vol. 4 (6), p. 341-351, 2008.
- Fiammetti R. *Le langage émotionnel du corps.* Dervy-Livres, 2004.
- Janov A. *The Biology of love.* New York: Prometheus Books, 2000.
- Kandel E. *In Search of Memory.* Norton, 2006. French translation, Paris: Odile Jacob 2007.
- Kim, J. *Physicalism, or something near enough.* Princeton University Press, 2005.
- Leopold D.A. and Logothetis N.K. Activity changes in early visual cortex reflects monkeys' percepts during binocular rivalry. *Nature 379:* 549-553, 1996.
- Lestienne R. *Miroirs et tiroirs de l'âme.* CNRS Editions, 2008.
- Radin D. *Entangled Minds: Extrasensory Experiences in a Quantum Reality.* Simon and Schuster, 2006.
- Reich W. *Character Analysis.* New York: Touchstone, 1972.
- Schrödinger, E. Discussion of Probability Relations between Separated Systems, *Proceedings of the Cambridge Philosophical Society* (31): 555–563, (1935).
- Tittel W., Brendel J., Gisin B., Herzog T.; Zbinden H. and Gisin N. (1998) Experimental demonstration of quantum-correlations over more than 10 kilometers, *Physical Review A, 57.*

- Tononi G. The Information Integration Theory of Consciousness, in *The Blackwell Companion To Consciousness.* Edited by Max Velmans and Susan Schneider, 2007.
- Uzan, P. *Conscience et physique quantique.* Paris: Vrin, 2013.
- Uzan P. On the Nature of Psychophysical Correlations *Mind and Matter* 12 (1), 2014a.
- Uzan P. The Quantum-like Approach to Psychosomatic Phenomena in Application, *Axiomathes* 25 (3), 2014b.
- Uzan P. Complementarity in Psychophysics, in *Lecture Notes in Computer Science,* Ed. Filk and Atmanspacher, Springer, Vol. 9535: 168-178, 2016.

About the Book

This book suggests a radical departure in approaching the mind-body problem. Instead of trying to causally relate subjective experience to the functioning of the body, it begins with the notion of the psychosomatic unity of the individual and looks for its conditions of possibility. This text shows that what makes this unity possible is the generalized entanglement relation that connects a person's subjective experience with its body functioning in a specific way.

In addition to providing a significant contribution to the long-standing philosophical debate about the nature of the mind-body connection, this change of perspective based on the concept of generalized entanglement allows for exploring a holistic approach to health. It can for example explain the existence of body memory and leads to a better understanding of the genesis and evolution of internal diseases, allowing for the development of mind-body therapies. This volume also provides new insights into mental disorders and sets the theoretical basis of self-healing methods appealing to students, researchers and professionals in the fields.

Contents

Chapter 1
Mind-Body Connection and Causation

This chapter deals with the possibility of interpreting the mind-body connection as a causal relationship. First note that this question, like any survey on the nature of the mind-body connection, has not always been thought as legitimate within the framework of science. For example, the behaviourist current developed at the end of the nineteenth century considered the mind as an impenetrable black box, which means that only what enters (stimuli) and leaves (behaviour) could be the subject of a rigorous scientific study, thus carefully avoiding any reference to the existence of mental states and to their content (Watson 1925; Skinner 1969). Consequently, according to this position, the only relationship that science could establish is the one that links the stimulus to the behavioural reaction, but not between the stimulus and the mind. The question of the correlation between bodily processes and subjective life was thus excluded from the domain of experimental science and left to the domain of metaphysics. However, advances in cognitive psychology, psychosomatic and, more recently, psycho-neuro-immuno-endocrinology show the refusal to take the mental states contents into account as an anti-scientific attitude.

Another philosophical position refusing to probe further the nature of the correlations between body and mind can be traced back to Hume's empirism (Hume 1740). For Hume, our concept of causation *is derived from experience:* it is the habit we take in anticipating or inferring one event from another, following the repeated consecutiveness of these two events in experience. Along this line of thought, causation is merely reduced to a regular association between cause and effect, the "cause" being contiguous with its immediate effects and preceding them. Causation would thus be nothing more than the observation of such an uniformity in the feeling of conjunction between two series of events. As a consequence of Hume's concept of causation it would be absurd to wonder whether our mental and our body could "influence" each other or, more generally, to investigate the nature of the mind-body correlations since their only observation is sufficient to establish that mental and bodily events are linked by causation. Hume's empiricist concept of causation thus bans any further investigation into the nature of this connection. If the weakness of

P. Uzan, *Mind-Body Entanglement*, SpringerBriefs in Philosophy,
https://doi.org/10.1007/978-3-030-90173-8_1

Hume's concept of causation has since been demonstrated on general grounds (Beebee 2016), this weakness has even more dire consequences when dealing with the mind-body connection because it grounds the concept of mental causation on our only experience which can be misleading, as explained below. Before exploring the main attempts to interpret the mind-body connection as a causal relation, this essential distinction will first be explained on experimental grounds.

1.1 Which Concept of "Causation" Are We Referring To?

We are naturally inclined to think that our conscious intentions can be causally responsible for our actions, referring for example to the fact that our conscious decision to raise an arm is followed by the realization of this movement, which is the expected effect. However, among the diversity of concepts of causation that have been proposed and that will be presented hereafter, which concept of causation are we referring to in order to characterize this "causal responsibility" of our conscious intentions?

The concept of causation plays a central role in everyday life and is a touchstone for explaining phenomena in science. The search for cause(s) of their appearance, which indeed relies on the paradigm of the moment, is generally regarded as the foundation of their rational study. However, the question of building a concept of causation susceptible to be applied to the mind-body connection has given rise to many discussions (Descartes 1664; Chalmers 1996; Kim 2005; Esfeld 2005). Before analysing the main attempts to interpret causally the mind-body connection, let us make an essential distinction between the *feeling* that our conscious intentions can be fully responsible for our actions and their *real* causal responsibility.

Libet's experiments (Libet 1985) have shown that when a subject takes the conscious decision to move their body the motor areas of their brain susceptible to trigger this movement have *already* been activated. This activation of the brain, signaled by a "readiness potential", occurs about 350 ms *before* the conscious decision. These experiments seem to show that the conscious decision would only be an epiphenomenal, psychological manifestation of a previous unconscious brain activity *that actually decides to perform the movement*. Generally speaking, they seem to undermine the idea that our conscious intentions could really cause our acts, and even the very existence of free will: How can I freely decide something if my brain has already begun to act in the moment I am consciously choosing?

These results have been extensively discussed, regarding in particular the reliability of the moment the subject reports their conscious intention (Gomes 1998). One can for example mention William James' notion of "specious present" that assigns a kind of thickness to the present and could then mislead our perceptions of the moment of the conscious decision. However, a recent experiment performed by Soon and collaborators has confirmed and even refined these results (Soon et al. 2008). In this experiment, the subject can choose between pushing a button with his left hand or another one with his right hand and decides on the moment to act.

During the experiment, the subject's brain is observed by MRI, which allows to simultaneously analyze the activity of diverse brain areas. MRI imaging has shown us that preparatory brain activity exists *7–10 s before* the subject makes a decision. Furthermore, this brain activity makes possible to predict, to a certain extent, *which side the subject will push the button*. These results therefore seem to call into question even more radically the idea that our conscious intentions could really be "causally responsible" for our acts.

This line of thought has been explored by psychologist Daniel Wegner who has confirmed on an experimental basis that *our sense of free will does not reflect the underlying causes of our behaviours* (Wegner and Wheatley 1999). In one of these experiments a subject is placed in front of mirrors positioned in a way that leads the subject to believe that what appears to be their arm is in fact someone else's. The subject then receives instructions to make certain arm movements and, immediately afterwards, the arm that the subject sees through the mirror scheme executes each of these movements. The subject then reports the impression of being at the origin of this illusory movement seen in the mirror. Wegner analyses these kinds of experiments as follows:

> The experience of willing an act arises from interpreting one's thought as the cause of the act. Conscious will is thus experienced as a function of the priority, consistency, and exclusivity of the thought about the action. The thought must occur before the action, be consistent with the action and not be accompanied by other causes.[1]

Libet's experiments, their extensions by Soon and team and by Wegner thus seem to clearly establish *a distinction between our feeling of causal efficacy of our conscious will and the supposed real causal connection between our thoughts and our actions*. As pointed by Wegner (1999, p. 480):

> …it may be that people experience conscious will when they interpret their own thought as the cause of their action. This idea means that people can experience conscious will quite independent of any actual causal connection between their thoughts and actions.

This essential distinction between the *feeling* that our conscious intentions can be fully responsible for our actions and their *actual* responsibility, which may even be completely independent, shows that the generally uncontested idea of a causal efficacy of the mind over the body and the world might be only a simple *illusion*. Keeping this conclusion in mind, let us now focus on the various attempts to causally interpret the mind-body connection *without merely mentioning our feeling that our thoughts are causally responsible for our actions*.

In order to illustrate the main causal approaches developed, consider the phenomenon that an apple coming off the branch of a tree falls to the ground. The notion of "cause" used to explain this phenomenon is relative to the paradigm of the moment: according to the Aristotelian paradigm (Aristotle's Physics), this

---

[1] Wegner has shown on an experimental basis that the feeling of causal responsibility of our conscious decision is generated by the satisfaction of these three conditions: its temporal *anteriority* on the action (but not a too long one), its *consistency* with it and the apparent *exclusion* of any other potential cause of it (Wegner 1999).

phenomenon is explained by saying that the earth is the "natural" place of sub-lunar objects (Aristotle 350 BCE), which is thus regarded as the cause of its fall; whereas, according to classical mechanics, the cause of this phenomenon is that the earth exerts a gravitational attraction on the apple. There is a profound difference between these two explanations of the same phenomenon: in classical mechanic the gravitational attraction of the earth on the apple is *a mechanical action* that is applied on it and gives rise to its motion whose characteristics (trajectory, velocity, acceleration) can be calculated at each moment from the physical laws–namely, Newton's law of universal attraction and the second principle of dynamics. The notion of causation is here defined in terms of *a mechanical process* relating the gravitational field generated by the earth and the motion of the apple via the gravitational force undergone by it. This mechanical process can be described as a succession of events taking place in space and giving rise to an exchange of forces, impulses and energy between the earth and the apple: generation of a gravitational field by the earth (the "cause"), propagation of this field to the apple, which is a massive object, and generation of a gravitational force capable to move it (the "effect").[2] In contrast, within the Aristotelian paradigm the explanation of this phenomenon *does not mention any mechanical process* that would explain the motion of the apple. This approach only refers to a "final cause" that would explain the motion of the apple towards its "natural" place, the earth. The notion of cause is thus here merely identified to that of *explanation* within the Aristotelian paradigm and does not rely on the existence of any mechanism or any succession of events taking place in space. These two possible explanations of the fall of the apple to the ground illustrate the two main approaches to causation when considering *two single events* c and e, respectively regarded as the "cause" and the "effect":

1. The *processual approach* of causation defines causation as the existence of a succession of events that transmit a property susceptible to producing the considered "effect" e from the "cause" c (Reichenbach 1956; Salmon 1984). It obviously applies in physics since it refers to the precise mechanical process that involves a transfer of matter, energy or information satisfying the relativistic constraints (Kistler 1998). However, as will be clarified hereafter, the processual approach of causation is highly problematic for interpreting the mind-body connection. This will be shown hereafter from conceptual/logical considerations, regarding Descartes' concept of mind-body interactionism and also the "non-

[2] Note that a definition of the notion of "mechanism" intended to be more general has been proposed by Illari and Jon (2012). It defines the notion of mechanism as follows: "A mechanism for a phenomenon consists of entities and activities organized in such a way that they are responsible for the phenomenon". However, the problem with this definition is that the notion of "responsibility" it refers to …refers in turn either to the alleged existence of a "mechanism", which makes it circular, or to that of *causal* responsibility, which makes it rather vague and without any specific significance if we account for the great diversity of the approaches to causation (see above). We thus prefer to keep a definition of mechanism that directly relies on the existence of *a mechanical process* in the original sense of mechanics, that is, a process that involves a succession of events taking place in space and involving exchanges of forces, impulses and energy.

reductionist" physicalist approaches to mental causation. Moreover, as will be explained in Chap. 4, the possibility of mind-body "signaling", that is, the possible existence of causal interactions between mental and somatic features in a person, irrespective of their particular form, can be rejected *experimentally* in as much as its measurable consequences are not satisfied.

2. The *explanatory approach* to causation interprets this notion by that of explanation–an approach developed by logical empiricism (Hempel 1965, pp. 245–290). In contrast with the previous processual approach, the explanatory interpretation does not specifically refer to any causal chain from a "cause"-event to the phenomenon in consideration, which would be its "effect". As a consequence, this approach does not say anything about the yet expected anteriority of the "cause" on the "effect" and could then allow self-causation. For example, in the previous Aristotelian explanation of the apple's fall to the ground, the "cause", the fact according to which the earth is the "natural" place of sub-lunar objects, and the "effect", the fall of the apple, are not a priori linked by any temporal order. Moreover, interpreting causation by referring to explanations of phenomena which are necessarily provided *within the paradigm of the moment* cannot grasp the *specificity* of the concept of causation, and then deprives it of stability and robustness. This specificity is to really generate or produce, independently of any paradigm, an "effect" from a "cause", for example by qualifying causation in terms of "generation, or effective production and determination" (Kim 2005).

   Now considering the connection between *series of events*, the notion of causation connecting a set C of "causes"-events and a set E of "effects"-events can be characterized in probabilistic terms or by referring to the notion of agentivity:

3. The *probabilistic approach* (Eells 1991) according to which an event c of C is the cause of an event e of E if the conditional probability of occurrence of e when c is realized is greater than the conditional probability of occurrence of e when c is not realized, which can be written as: $p(e/c) > p(e/\rceil c)$. Along the same idea, the *counterfactual approach* to causation (Lewis 1986) asserts that c is cause of e at the necessary and sufficient condition that if c did not happen e would not have occurred. It can then be considered as a limiting case of the probabilistic approach of causation where $p(e/\rceil c) = 0$ and then $p(e/c) > 0$. As is the case for the probabilistic approach, the counterfactual approach interprets the concept of causation in terms of correlations between the two considered series of events. The possible application of the counterfactual interpretation of causation for dealing with the mind-body connection will be analyzed below in this chapter.
4. The *agentive approach* developed by Gasking (1955) and by Price (1992), according to which the "causes" of the set C are the events that can be used as means to obtain the "effects" of the set E. Along the same line of thought, Woodward (2003) has developed the so-called *interventionist approach* of cau-

sation regarding *variables*: a variable[3] X can be considered as a “cause” of a variable Y with respect to a set of variables V if and only if an intervention on X that leaves unchanged all other possible variables in V that are not on the directed path from X to Y induces a change in Y (Woodward 2003, p. 90). Woodward’s interventionist account of causation and its use for interpreting the notion of mental causation will also be analysed below in this chapter.

Could one or several of these approaches of causation be used to characterize the mind-body connection? The following of this chapter will explain the difficulty of interpreting psychophysical correlations as causal relations. We will first remind the reader of the difficulties faced by the traditional responses to this problem, these “traditional” responses being respectively developed according to Descartes’ dualism of substances and its recent extensions, and according to the dualism of properties within a physicalist framework-the so-called “non-reductive physicalism”. Two recent suggestions aiming to restore a concept of mental-to-physical causation will then be analysed. These two proposals respectively appeal to Woodward’s interventionist account of causation and to Lewis’ counterfactual approach to causation.

1.2 Dualism and Causal Efficacy

The question about the nature of psychophysical correlations first refers to the distinction of what we perceive as “mental” and what we perceive as “physical”. Do the terms “psycho” and “physical” refer (or not) to two *distinct* entities, which can be two distinct substances or two distinct aspects of a same person? If we refer to our experience, it seems that the mental states, like thoughts, desires, feelings and sensations have properties that are seemingly not those of our body. We have a private access to mental states since they can only be reported by the subject who is experiencing them; they can be intentional (about something) and give meaning to our actions and to the world we live in, and moreover can be accompanied by a conscious experience (Nagel 1974; Chalmers 2007). In contrast, physical states can be described publicly, in the third-person language of science, they do not have any particular meaning for a subject and are not characterized subjectively, in terms of qualia. Let us set a first important principle, called *Irreducibility*, whose validity will be discussed hereafter. This principle asserts that mental properties are not reducible to, and not identical with, physical (bodily) properties.

On the other hand, the causal efficacy of the mental on the physical, as well as the causal efficacy of the physical on the mental, *seem* fully justified. Downward causation seems to be necessary to explain the possibility of voluntary action -for example, that the desire to raise one’s arm can be actualised- and that of human agency, which allows us to act in the world. Upward causation also seems necessary in

[3]The notion of “variable” is more general than that of “property” since a variable can take several values that define corresponding properties.

explaining that a physical event or a physical property can have mental effects -for example, that the chemical properties of an ingested drug can ease a mental distress. We are then intuitively inclined to accept a principle of *Causal efficacy of the mental on the physical* and, conversely, a principle of *Causal efficacy of the physical on the mental.*

However, the irreducibility of the mental to the physical seems impossible to reconcile with both downward and upward causation. To see this, two cases must be distinguished according to the interpretation that is given to the principle of *Irreducibility*.

1.2.1 Dualism of Substances: Elusive Psychophysical "Interactions"

René Descartes can be considered as the father of modern dualism and of the modern formulation of the mind-body problem we still refer to. According to Cartesian dualism, matter and mind are distinct, independent substances, respectively called "*res extensa*" by referring to the property of material things to be extended in space and "*res cogitans*", the thinking thing which is not extended in space. The body would be "a machine that wakes of itself", while the mind would be "an immaterial principle whose essence is only to think and which needs no place" (Descartes 1641). This modern formulation of dualism of substances finds its roots in Plato's distinction between things and ideas, respectively referring to the sensible and the intelligible worlds, in contrast with pre-Socratic philosophy, which generally was in search of a unique principle of existence. The possibility of voluntary bodily motions was explained by Descartes in terms of *causal interaction* between the mind and the brain (Fig. 1.1). This interaction was supposed to take place in the pineal gland, an organ not supposed to exist in pair and located between the two hemispheres, just above the Sylvius aqueduct which was supposed to act on it and give rise to sensations.[4]

Irreducibility is thus interpreted here as a dualism of substances and we have to explain how the immaterial "thinking" substance, not extended in space, can have

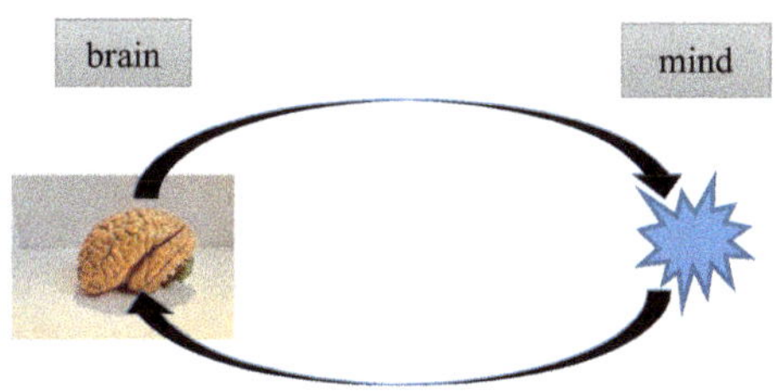

Fig. 1.1 Mind-brain interactionism

[4] Note that beyond the particular role of the pineal gland in Descartes' philosophy, the view according to which some brain parts, like the limbic system or the hypothalamus, would act as liaison between mental and visceral processes is still relevant in today's neurobiology and medicine.

any causal efficiency on matter, localised in space and reciprocally. As rightly asserted by Descartes, this alleged interaction between mind and body cannot be understood as genuinely physical, since the thinking substance is not supposed to be a physical object located in space. However, asserting that the mind-brain interaction would take place in an organ well-located in space, namely the pineal gland, actually suggests a kind of *spatialization of the mind* in contradiction with Descartes' claim. This idea has been explored by contemporary authors, like Lycan (2009),[5] in order to ensure a direct causal contact between the mind and the brain, thus making the existence of such a physical-type interaction plausible. Moreover, this solution would bring an answer to the "pairing problem" of explaining the pairing of a person's specific immaterial ego with their own body and not that of another person (Foster 1991; Kim 2005, Chap. 3).

Whether Descartes' thinking substance can be considered to be "spatialisable" or not, current neurobiology has undermined Descartes' assertions about the pineal gland and its alleged role in the mind-body interaction. It has been shown that the pineal gland is a conjugated organ (its two hemispheres being almost fused together) whose essential function is the regulation of biological rhythms by secreting melatonin (Pfeffer et al. 2017). However, this did not prevent the development of the interactionist solution by great figures of philosophy and science. Popper compared the action of the mental over the physical with a physical force of which it would have common properties. For example, its intentionality oriented towards something could be represented, like for a physical force, by a vector (Popper et al. 1993). Within the quantum paradigm, the interactionist position has been supported by appealing to the subjectivist interpretation of the wave function collapse suggested by von Neumann (1955) and Wigner (1983). According to this interpretation, the subject's "consciousness", "mind" or "faculty of introspection" would play an essential role in the collapse of the wave function by selecting one of the possible results of a measurement. To explain the brain-mind interaction, Beck and Eccles respectively refer to two types of elementary units: the physical "dendrons" and the mental "psychons." The clustering of the dendrites of the pyramidal cells ascending to the upper layer of the cortex define structural micro-unities of the cerebral cortex (Peters and Kara 1987), called "dendrons", which are present in large quantities in the human brain (40 million). The dendrons, which would then constitute the seat of the micro-neural events, are supposed to interact with the mind, that is, each dendron would be penetrated by an elementary mental unit or "psychon." In virtue of this brain-mind interaction, a mental intention could subtly change the probability distribution associated with exocytosis (the release of neurotransmitters) and consequently influence the transfer of information between neurons -a phenomenon that would be amplified to the macroscopic level in reason of the large number of synaptic clefts of the pyramidal neurons of the cortex (Eccles 1994).

---

[5]Lycan has suggested that minds would be located "in the head behind the eyes" (Lycan 2009, part III)

However, if it is more tenable to focus on the exocytosis phenomenon taking place in all chemical synapses of the brain rather than on the role of some particular organ, the nature of this "interaction" between the two heterogeneous mental and material substances still remains unclear. If physics accounts for physical-to-physical causation and psychology for mental-to-mental causation, what could be the laws governing psychophysical causation? Models of such psychophysical interaction have been provided but they have not been very convincing up to now. Walker has for example suggested evaluating the action of mind over matter by asserting that the amount of information processed by consciousness in a physical (quantum) transition is equal to Shannon's statistical information associated with the probability of this transition (Walker 1975). In order to increase the maximum psychokinetic "effect", Walker's suggestion has been refined by Mattuck (1980) by assuming that mind would be capable of performing a quantum transition in several steps. However, not only have these attempts given rise to controversial conclusions relative to the alleged powers of mind (Alcock 1989) but, as for Beck's and Eccles' model, they rely on a subjectivist interpretation of the wave function collapse which is far from being accepted by the scientific community for its lack of clarity and coherence (Uzan 2013). Furthermore, this interpretation seems rather useless in regard to other attempts of explaining the wave function collapse *within the framework of quantum mechanics*. It is for example the case of the decoherence approach proposed by Zurek (1981–1982) that does not appeal to any particular role of the observer's "consciousness" or "faculty of introspection". Let us also mention the fact that the interactionist models of the mind-matter connection seriously threaten the validity of the physical laws. The time-energy quantum indetermination relation can make acceptable a *momentary* violation of the principle of conservation of energy of a physical system due to an alleged action of the mind on this system. However, asserting with Eccles (1994) that the mental substance can change the probabilities of neurotransmitter release leads to the idea that Born's rule, an essential rule of quantum theory from which are computed these probabilities, is no longer valid.

Consequently, the idea of irreducibility of the mental to the physical, which is here interpreted by the dualism of substances, cannot be easily reconciled with the idea of psychophysical causation since it refers to a notion of "interaction" still struggling to find legitimacy.

1.2.2 Non-reductive Physicalism: Logical Inconsistency of Psychophysical Causation

If mental to physical irreducibility is now interpreted as a *dualism of properties*, we are prompted to ask ourselves: "what is the entity these mental and physical 'properties' refer to?" The mainstream answer to this question is to say that they are *properties of matter*, supposing that everything in our world derives a priori from matter. "Physicalism" is used here as a synonym of this latter position, namely *materialism,*[6] asserting that matter is the single, fundamental substance in nature and that all phenomena result from material interactions -which entails, in particular, that mental processes and subjective experience should be explained from brain processes. The physicalist (or materialist) position is combined here with the non-reductionist[7] view according to which *mental properties are nevertheless distinct from physical properties* -hence the name of "non-reductive" physicalism. However, is the distinction between mental and physical properties of matter really compatible with mental causation?

Before responding to this question on general grounds, let us notice that non-reductive physicalism has been implemented in the biological realm by the idea of *emergence* of mental states and mental processes from the brain activity. This means that brain processes would give rise or produce or "cause" mental processes and, in particular, subjective experience. At first sight this approach seems to lead to the idea according to which psychic life is a by-product or an epiphenomenon of brain processes, without any function or any possible "effect" on the body (Ribot 1896)- like the noise produced by a car is useless to its functioning. However, the very possibility of performing experiments[8] that detect the moment a subject becomes conscious of perceiving a stimulus and decides or not to *consciously* react seems to show that subjective experience is endowed with some autonomy, which then undermines the epiphenomenalism doctrine.

A more realistic version of emergentism that takes into account the autonomy of subjective experience has been upheld. According to this position which has been explored by Searle (1997), by Sperry (1983) and, more recently, by Lestienne (2012), the mental is conceived as an emergent, higher-level property of the brain that differs from its physical properties and would be governed by its own laws that transcend those governing the brain processes. In turn, as suggested by Sperry, the subjective properties would be capable of exerting "control over the biophysical and chemical activities at subordinate levels", thereby assigning some autonomy to consciousness. In order to explain the emergence of consciousness from the brain, Searle upholds the thesis of "biological naturalism": on the one hand, consciousness would be a biological phenomenon "caused" by brain processes in the most evolved of animals; on the other hand, this phenomenon would be "natural" like digestion, photosynthesis or mitosis. According to this emergentist position, mental phenomena and, in particular, consciousness could not be exclusively explained from the

[6]The term "physicalism" actually has a more specific meaning, as it states that all phenomena of nature are "physical", that is, they can be described by the laws of physics. However, this specificity does not have much relevance to the question debated here and will not be retained in the following, as in most of the work on this subject.

[7]As we will see below (Sect. 1.4), the *reductionist* form of physicalism leads to the problem of explaining the subjective experience (qualia) that accompanies the cerebral activity. The non-reductionist form of physicalism aims getting around this problem by maintaining a distinction between mental and physical properties.

[8]Any experiment about visual consciousness can be mentioned (Koch 2006).

individual properties of brain constituents or even from their combination, but should be understood as *global* properties of the brain whose emergence requires a minimal level of complexity (acquired in evolution).

However, if the latter emergentist approach aims avoiding reductionism by asserting that the mental should be understood as a new, emergent property of the brain, it is nevertheless faced with the question of the *explanatory gap* (Chalmers 1996; Levine 1983; Nagel 1974). How could a third person description, as precise as we want, of this cerebral activity by biophysical or biochemical laws "explain" *the subjective experience* that accompanies this activity? Subjective experience can be defined by "what it is like" to realize a cognitive act, like that of perceiving a red flower, or to perform an action, like drinking a coffee or raising an arm, thus referring to *the content of consciousness*. As explained by Noë and Thompson (2004), this content is relative to an ensemble of personal characteristics, like the subject's own sensibility, his beliefs, his desires and his own rationality, whereas physiological processes are of sub-personal nature and do not consequently rely on personal characteristics. How could a *description* of sub-personal, physiological processes in the third person explain the *private experience* of a particular subject? Assuming the primacy of matter can certainly allow a description of *neural correlates* of subjective experience; however, can such an objective description account for *the private content of subjective experience*?

On general grounds, without explicitly referring to the biological or physical nature of the relation between mental states and brain states, non-reductive physicalism is faced with Kim's causal argument exclusion (Kim 2005) whose radical conclusion is that this position is logically inconsistent: non-reductive physicalism *cannot be reconciled* with mental causation. In order to see this, let us clearly notice the two principles on which physicalism is grounded:

Physicalism first supposes that:

(a) *Mental properties supervene on physical properties.*

The supervenience of the mental on the physical asserts that mental states are *grounded in the physical*, which means that their very existence depends on, and is even determined by some physical properties, called their physical « realizers ». The supervenience of the mental on the physical, illustrated by Fig. 1.2, ensures that the mental properties of matter are rooted in the physical realm and cannot then be designed independently. The relation of supervenience, which is *a priori* not supposed to be an identity or a causal relation, can be characterized very generally by the following property:

> …a set of properties A supervenes upon another set B just in case no two things can differ with respect to A-properties without also differing with respect to their B-properties. (Mc Laughlin and Bennett 2011).

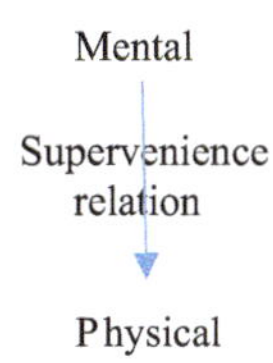

Fig. 1.2 Supervenience of the mental on the physical

The second principle underlying physicalism is:

(b) *The causal closure of the physical world*

Broadly understood, this principle asserts that the events that are causally related to a physical event are also physical events. In its strongest version, this principle reads as:

(b1) All the possible causes *and* all the possible consequences of a physical event *are necessarily of physical order* (Fig. 1.3).

Fig. 1.3 Causal closure of the physical world

This strong version of the "causal closure" of the physical world is proposed here in full agreement with the idea of closure of a domain or a set by a relation. It thus strictly expresses the physicalist position. However, weaker versions of this principle have been proposed. Generally, philosophers adopt versions of this principle that only focus on the possible causes of a physical event (and not on its possible consequences), as is the case for Kim (1993, p. 280):

(b2) No physical event has a cause outside the physical domain, a version which is indeed sufficient for deriving his "exclusion argument" that shows the absence of any causal efficacy of the mental on the physical (see hereafter). An even weaker version of this principle, called "Completeness of the Physical" by some authors (Crane 2001; Papineau 1993), states that any physical event has an only *sufficient* physical cause but *do not preclude a priori* the existence of a non-physical (mental) cause:

(b3) Any physical effect has a sufficient physical cause. The latter version of the principle of causal closure of the physical world leads to the possibility of *overdetermination* of a physical property by a mental property and has been upheld by some philosophers in order to save the idea of mental causation (Mills 1996). However, if Kim's weaker version (b.2) genuinely expresses the physicalist position regarding the notion of cause, we cannot say the same of the even weaker version (b.3). For the existence of a non-physical event that would "cause" a physical phenomenon is completely irrelevant of physicalism, namely the idea that our world derives from matter and that any (material) phenomenon can ultimately be explained by material interactions. We cannot at the same time assume that the world is made up of *purely physical entities* governed by physical laws (as required by physicalism) and accept that non-physical entities be responsible for its functioning. The idea of overdetermination of physical events by mental events being inappropriate within a strict *physicalist* position it will not be further considered in this section.

Taking these two principles into account, namely the supervenience of the mental on the physical and the causal closure of the physical world, it is then possible to show that within non-reductive physicalism, the dualism of properties is *logically incompatible* with both downward and upward causation. The incompatibility with upward causation is obvious by the Causal closure of the physical domain (which asserts that the possible consequences of a physical event are necessary of physical order), while the incompatibility of Irreductibility with downward causation, shown by Kim's exclusion argument, needs more clarifications. These clarifications are provided by referring to the following diagram (Fig. 1.4):

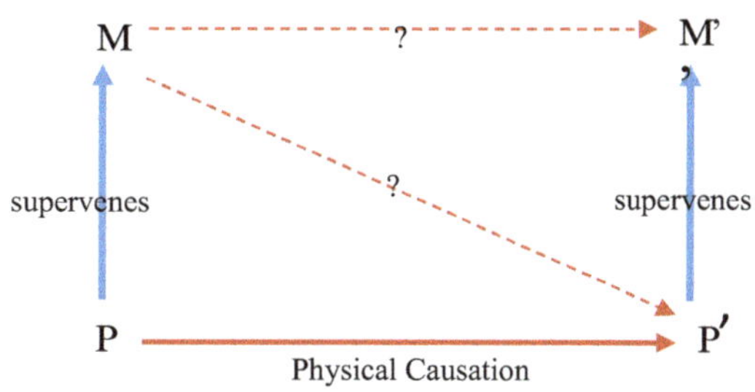

Fig. 1.4 About mental causation

Suppose that the mental property M causes the physical property P′ (see the diagram). Then, by the Causal closure of the physical domain (in Kim's version (b.2)), the *only possible (set of) cause(s)* of P′ is a physical (set of) cause(s), which is here denoted as P. Consequently, this physical cause P *pre-empts* M, which then has *no causal power on P′*. This is Kim's exclusion argument.

Moreover, non-reductionist physicalism also shows the impossibility of the mental-to-mental causation. Let us suppose that M causes M'. Then, by the Supervenience of the mental on the physical (a), M' exists in virtue of its physical realizer P′, and then M must cause P′ for M' to exist. Since, by the previous case, M has no causal power on P′, *M has neither causal power on M' and then the mental has no causal efficacy at all.*

Thus, due to Kim's exclusion argument, non-reductive physicalism, namely the conjunction of dualism of properties and physicalism, cannot lead to a coherent causal interpretation of the mind-body connection. Could Kim's exclusion argument be overcome?

1.3 The Hopeless Attempts to Build a Concept of Mental Causation

Several attempts aiming to build a concept of mental-to-physical causation that overcomes Kim's exclusion argument have been developed. These attempts have first tried to work out a specific definition of the notion of mental "cause" according two directions. The "autonomy solution", developed by Jackson (1996), is based on the idea that the "causal" explanations of living beings' behavior offered by psychology can be totally independent from those that physical sciences could provide, for example by explaining their bodily movement. This autonomy of psychological explanations, which refer to their own ontology and to their specific laws, independent from those offered by "lower-level" physical sciences, would thus allow us to conceive a notion of "mental cause" that does not compete with that of physical cause. Since this notion of mental cause does not appeal to a physical implementation, the causal efficacy of mental properties cannot be excluded any more. According to this "autonomy solution", both physical (P) and mental (M) causes, which have different and independent statutes, would then coexist. The second proposal aiming to refute the exclusion argument is based on the opposite idea that mental properties are strongly dependent on their physical realizers since they "inherit" their causal power–a solution which has been, for example, developed by Shoemaker (2001). Like in the previous case, but from an opposite point of view, physical causation does not exclude mental causation since there is no concurrence between them but a perfect cooperation of these two qualitatively different "causes." Both approaches have tried to work out a definition of the mental-to-physical causation that does not compete with the physical notion of causation and then would not be excluded by the existence of physical realizers of any mental properties.

These rather imprecise ideas of "autonomy" of mental properties and "inheritance" of their physical realizers have been captured and implemented in two recent proposals on which we will now exclusively focus. These two proposals, qualified by Butterfield of "rebuttals" of Kim's exclusion argument (Butterfield 2012), are based on more precise characterizations of the concept of "causation". The first one has been formulated within Woodward's interventionist account of causation (Woodward 2003). It has been presented by Shapiro and Sober (2007) and developed in further detail by Woodward (2011) himself in his response to Baumgartner's critique (2010). The second is developed within the counterfactual account of causation and has been developed by List and Menzies (2009).

1.3.1 Woodward's Interventionist Account of Causation and Mental Causation

It will not be necessary to develop the different aspects of Woodward's proposal here (for example, regarding the distinction between type level causation and singular causation or that between direct cause and contributing cause) but we will consider its general idea, which is sufficient for our purpose. According to Woodward, a variable[9] X can be considered as a "cause" of a variable Y with respect to a set of variables V if and only if an intervention on X that leaves unchanged all other

[9]As mentioned above in the general presentation of the agentive approach to causation, the notion of "variable" is more general than that of "property" since a variable can take several values that define corresponding properties.

possible variables in V that are not on the directed path from X to Y induces a change in Y (Woodward 2003, p. 90).[10] An "intervention" on X is defined as a cause of X that satisfies several conditions (p. 98), like that according to which this cause must be "statistically independent of any variable Z that causes Y and that is on a directed path that does not go through X."

Is it possible to overcome Kim's exclusion argument and to provide a clear meaning of the concept of "mental causation" within Woodward's interventionist account of causation? Our answer to this question is negative.

As pointed out by Baumgartner (2010), the interventionist account of causation cannot refute Kim's exclusion argument for the following reason. In order to show that M is a cause of P′ (see Fig. 1.4 above) we would have to show that an intervention on M that leaves P unchanged will change P′ (P being a physical cause of P′ which is not on the path M-P′, as required in the definition above). However, by the very definition of the supervenience relation (see above), if M supervenes on P *it cannot be changed without changing its supervenience base P.* This makes the interventionist account of causation inapplicable to interpret the notion of mental causation.

In order to overcome this difficulty, Shapiro and Sober (2007), as well as Woodward (2011), have provided arguments to justify that the interventionist account of causation *must not* be applied to the supervenience basis of mental states. According to Shapiro and Sober (2007, p. 11), M must be conceived of as a *functional property*, that is, a second-order property "of having some property that fills some functional role" and cannot then possess "any powers beyond those of a first-order property", in particular beyond those of P. This argument would then refute Kim's exclusion argument, which relies on the idea that M has no additional causal power on P′ than that of P. Woodward also states that it would be "inappropriate to control for supervenience bases in asserting the causal efficacy of supervening properties" and suggests that an assessment of mental causation could be obtained by controlling for "other causes" than supervenience bases of mental events (Woodward 2011, p. 13).

However, Shapiro's and Sober's argument, as well as Woodward's response to Baumgartner do not only remove any specific meaning from the concept of "mental causation" but also suffer from logical inconsistencies:

1. On general grounds, several powerful arguments have already been provided to show the inadequacy of Woodward's interventionist account of causation:
    (a) *Cicularity.* As noted by Strevens (2008, p. 15), Woodward defines the notion of "causation" (Woodward 2003, p. 59) by appealing to that of "intervention". However, the definition of the notion of "intervention" (p. 90) uses

[10] Note that Shapiro and Sober (2007) have reformulated this definition by requiring that this intervention must hold unchanged all possible *common* causes of X and Y: "X causes Y" if "wiggling X while holding fixed whatever *common causes* there may be to X and Y will be associated with a change in Y" (p. 6).

that of "causation", which makes this approach confusing. In the end, we can say that nothing is precisely defined.

(b) *Nomological dependence rather than "causation"*. As noted by Kistler (2013), Woodward's interventionist account of causation characterises the *nomological dependence of properties* related by a law of functional dependence rather than their causal relation. This nomological dependence is similar to that of the quantities involved in a physical law, whose variations are related by the law but are not necessarily related by a causal relation -in particular, their changes occur at the same time in contradiction with the temporal succession inherent to a causal relation.[11]

2. More specifically, by attempting to refute Kim's exclusion argument within this interventionist account of causation, these authors assert that it would be inappropriate to control the physical supervenience basis of mental properties. For Shapiro and Sober, mental properties are considered to be functional, second-order, properties *that do not (and should not) possess causal powers beyond those of their supervenience basis* (Shapiro and Sober 2007, p. 11). We can then ask these authors which kind of "causal" power they assign to mental properties. How is characterized mental *causation* here? If physical causation is clearly defined by (relativistic) physics, it seems that the "mental causation" is only defined here *by its lack of causal power!* The least we can say is that such a concept of mental causation completely lacks of consistency.[12]

Along this line of thought, Woodward has asserted that it would be a methodological mistake to control supervenience bases when testing the causal efficacy of variables representing mental properties (Woodward 2011, p. 29). He has suggested to weaken the initial interventionist account of causation in order to allow an intervention variable for X with respect to Y to be causally connected to Y through the supervenience basis of X. This means that an intervention on the mental variable M with respect to the physical variable P′ could induce changes on the supervenience basis P of M (Woodward 2011, p. 27). Or, more explicitly, that a change in the mental property M can still be regarded as an "intervention" that causes a change in the physical property P′ even though its supervenience basis P cannot remain unchanged. However, if this weakening of the initial interventionist account of causation makes the interventionist proposal compatible with the non-reductionist physicalism, it is not clear at all that a change in M be the genuine "cause" of the change of P′. If a correlative change of its supervenience basis P is allowed, the latter can still be

[11] The concept of causation is generally associated with a temporal order, which means that the "cause" is supposed to occur before the "effect", at least in classical physics and for common sense.

[12] Along this line of thought, let us also mention Roy's proposal (Roy 2004) according to which mental and physical properties are related by a relation of "ontological dependence", which is supposed to clarify (or even to explain) the concept of mental causation. However, the concept of "ontological dependence" is far too weak to shed any light on the specific nature of the psycho-physical correlations. For example, we could say that anything in nature can "depend ontologically" on other things because they have a common origin according to the big bang theory.

taken as the exclusive cause of the change in P′, which works against the "non-reductive" physicalism thesis. The concept of "mental causation" cannot obviously be saved if the (causal) role of the supervenience, physical basis of the mental variables is not neutralized -like any experiment aimed to show the role of one of the parameters must neutralize the action of the other ones.

Thus, Kim's exclusion argument cannot be considered to be refuted by Shapiro's and Sober's argument or by Woodward's response to Baumgartner. This attempt to build a concept of "mental causation" by referring to Woodward's interventionist account of causation only emphasizes the inapplicability of this approach, already questionable for itself, to the specific question of the mental-to-physical causation.

1.3.2 The Counterfactual Account of Causation and Mental Causation

Another alleged "rebuttal" of Kim's exclusion argument has been proposed by List and Menzies (2009), based on the idea of "proportionality of causation" (p. 5). It ensures that "causes are specific enough for their effects, but no more specific than needed" (p. 5). This idea aims to replace the restrictive sense of the notion of causation as "generation, or effective production and determination" (in Kim's words) by a counterfactual-type account of causation (see Lewis 1986). More precisely, the idea of proportionality of causation is expressed as follows:

(C) F is said a "cause" of G in a sense that the presence of F makes a difference to the presence of G in the actual world, that is, "if and only if it is true in the actual world that (i) if F is present, G is necessary present; and (ii) if F is absent, G is necessary absent." (List and Menzies 2009, p. 6)

List and Menzies assert that the latter account of causation can be satisfied by the downward causation, which would then show that Kim's exclusion argument is false (p. 10, with notations adapted to the previous ones):

> Even when some [physical] property P is causally sufficient for some [physical] effect P′, a [mental] property M that supervenes on P can nonetheless be a cause of P′.

According to these authors, the mental property M (for example, a decision of raising an arm) would be *the genuine cause* of P′ (the raising of the arm) while its physical realizer P could be replaced by another physical property, say P1, *that could also be a sufficient cause of P′*. M and P′ would then always satisfy the condition (C) since the presence of M entails that of P′ and its absence entails that of P′, while it is not the case for P and P′ since P′ can still be actualised in the absence of P, with P1 being a sufficient cause of P′. The relation between M and P′ is qualified of "realisation-insensitive" by these authors (p. 16).

However, this attempt to refute Kim's exclusion argument can be questioned as follows:

1. As has been well known for a long time, the counterfactual account of causation can only capture the general idea of *correlation*. Examples can be provided in different fields (such as medicine, social and natural sciences) to show that the existence of counterfactual relations between events *does not imply* that of a causal connection between them. These examples refer to the possible existence of a common cause for two correlated series of events (like in the well-known example of the correlation between the amount of coffee drunk and the chance of cancer, which can both be caused by the stress level).[13] Moreover, in some cases such as EPR-like experiments in quantum physics, correlations satisfying the above counterfactual conditions neither can be explained by the existence of a causal interaction nor by that of a common cause.[14]
2. Beyond this general critique of the counterfactual account of causation, there is a more specific problem in List's and Menzies' argument. As in the case of the interventionist approach previously analysed, List's and Menzies' argument suffers from a lack of consistency. These authors have elegantly presented and justified the idea of proportionality of causation by enlightening examples, such as the one relative to the "good" dose of drug which is required for recovering from an illness (List and Menzies 2009, p. 5). This example clearly illustrates the fact that the condition (C) rules out the causes (the amounts of drug) that are either not specific enough or too specific to account for the change of the effect variable (the binary variable whose values are "recovery" or "non-recovery"). However, the problem is that the application of condition (C) to refute Kim's exclusion argument (p. 10) is unfounded. The example (p. 9–10) that shows that the condition (C) applies to the intention of a monkey of performing an action and not to the neural basis of this intention, which would only be a sufficient cause, is unjustified. If the idea of "proportionality of causation" legitimately applies to the case of the amount of drugs necessary for healing, why and how exactly would this idea be applied to the mental properties that are not supposed to be reducible to physical properties? No clear argument is provided to justify that the intention of the monkey is the *genuine* cause of his action, and not its physical cause. Actually, this assertion is only *postulated.*[15] Thus, it seems that List's and Menzies' argument, based on the notion of "proportionality of causation", also fails to refute Kim's exclusion argument.

Let us sum up the previous results. On one hand, *dualism of substances* cannot be easily reconciled with the idea of psychophysical causation, since it appeals to an elusive notion of psychophysical "interaction". On the other, non-reductionist

[13] See also Kistler's synthetic article (Kistler 2004).

[14] This feature of EPR-like correlations is qualified as "non-locality" (d'Espagnat 1994; Clauser et al. 1969). It will be explained in more detail in Chaps. 3 and 4, where it is used to build the concept of mind-body entanglement.

[15] Moreover, it can be noticed that this postulate is stronger than the assertion they want to demonstrate, namely, that a mental property can be a cause (and not *the genuine cause)* of a physical property in spite of the causal closure of the physical world.

physicalism, which relies on the *dualism of properties* of a single *material* substance, suggests concepts of psychophysical causation that crucially lack specificity and suffer from logical inconsistencies. Within a non-reductionist framework keeping the principle of Irreducibility of the mental to the physical, it consequently seems that psychophysical correlations cannot merely be conceived as direct causal connections, whatever their nature, between mental events and physical events. Should mental-to-physical causation be interpreted as *a physical-to-physical causation*, thus assuming that mental states could be "reduced" to physical states in a sense to be specified? The next section will explore this reductionist option of physicalism.

1.4 Reductionist Physicalism

The previously mentioned difficulty to reconcile mental causation and dualism has led us to ask whether mental-to-physical causation could be merely understood as a physical-to-physical causation, thus assuming that mental states could be "reduced" to physical states. This line of thought, called *reductionist physicalism*, deletes the principle of Irreducibility by asserting that mental processes are nothing but brain processes. It has been upheld by great philosophers, such as Kim (2005), and is currently the dominant position among neuroscientists, as asserted by Kandel, who worked on the brain substrates of memory:

> ...consciousness is a biological process that will eventually be explained in terms of molecular signaling pathways used by interacting populations of nerve cells... (Kandel 2007, p. 9).

or by Crick, who discovered with Watson the structure of DNA:

> ... "you", your joys and sorrows, your memories and your ambitions, the meaning you have of your personal identity and your free will, are not in fact, nothing more than the behaviour of a vast assembly of nerve cells as well as molecules associated with them (Crick 1994).

Along this line of thought, but from the viewpoint of cognitive science and artificial intelligence, let us also mention Minsky's claim that:

> Minds are simply what brains do (Minsky 1988a, 1988b).

Reductionist physicalism is pushed to the extreme by Dennett (1991) who contends that subjective experience would be nothing but an *illusion* produced by the integration of sensori-motor brain information processes and maintained by culture. Dennett, who, along with Hofstadter, is a supporter of the "strong" version of artificial intelligence, proposes to reduce consciousness to a computer program implemented on the brain-machine running in parallel -a machine which, moreover, would not have been particularly designed for this purpose (Hofstadter and Dennett 1987). According to this thesis, there would be no difference between a "conscious" being (according to Dennett) and a machine that behaves in the exact same way. We would be philosophical zombies, just like any other entity capable of perfectly imitating our behaviour without phenomenal conscious states. Dennett's position could thus be described as "eliminativist" insofar as the phenomenal aspect of consciousness is simply denied. What a person believes to be subjectively experiencing, in a private way, would be nothing but a process of integration of information in his brain, which is describable in the third person.[16]

More precisely, this "reduction" of mental states to physical or neurophysiological states can be understood in two different ways. First, it can be understood as a mere *identification. Type identity* asserts the identification of types (or classes) of mental events with types (or classes) of physical events and has been developed according to various degrees (Feigl 1958; Smart 1959; Armstrong 1968). For instance, the class of mental pains would be identified with the class of neural mechanisms that neuroscience currently associates with this pain and which are synthetically referred to as "C-fiber activations" in philosophical literature. An important evolution in the applicability of the type identity theory has to be mentioned: Place, Feigl and Smart did not initially intend to apply this theory to all mental states but only to *non-intentional* mental states such as pain; while intentional states such as beliefs or desires were interpreted from a behavioural point of view, in terms of dispositions.[17] It is on Armstrong's suggestion that identity theory has been applied to all mental states. A second clarification to be made about this theory is that for these authors identification of sensations to brain processes is conceived as *contingent*[18] *and empirical*, just as the identification of the evening star with the morning star relies on the empirical discovery that their reference is the same -whereas it was previously assumed that they refer to distinct objects. Referring to Frege's analysis of the distinction between sense and reference (Frege 1892), these authors thus claim that sensations and brain processes have distinct meanings but can have the same neurophysiological reference.

An important argument that has been opposed to type identity theory is the actual *multiple realizability* of mental states. As emphasized by Putnam (1967), a same mental state (like "being hungry") may be possessed by creatures that differ physically, physiologically or even neurologically from human beings, which then questions the possibility of identifying mental and physical types of events. Nevertheless, a weaker form of mental-physical "identification" that takes the multiple realizability of mental states into account can still be upheld. It is the *occurrence (or token) identity* according to which every *occurrence* of a mental event is identified to some *occurrence* of a physical event[19] (Kim 1966; Davidson 1980). Occurrence identity theory underlies contemporaneous neurobiological reductionist approaches, like

[16]Along this line of thought, reductionist physicalism could go so far as to assert, as Churchland (1996) does, that the vocabulary of psychology would disappear and be replaced exclusively by that of neuroscience.

[17]The "disposition" of an object refers to its potentialities or to its power, such as the property to be soluble in a given solvent.

[18]Note that this notion of contingent identity has been severely criticized by Kripke (1972) who conceived of the identity relation as *a necessary relation.*

[19]This means that the mental-physical identification can only be asserted for a particular individual of a particular specie, and at a specific moment.

that of Dennett's mentioned above, or that of Crick's who claims that mental processes and, in particular, subjective experience, are nothing more than neural processes (Crick 1994).

The second proposition to reduce mental states to physical states is a so-called "functional reduction" and has been for example explored by Chalmers (1996), Kim (2005) or Levine (1983). The functional reduction of mental states first provides a causal definition of these states, which means that it exclusively focuses on the fact that they are caused by some inputs and, in turn, cause certain behaviours and other mental states. Then, by appealing to scientific knowledge, it tries to determinate their physical realizers, which are the physical properties or the physical processes that are capable of performing this causal work, and the way it is performed (Esfeld 2005, Chap. 14; Kim 2005, Chap. 6). In other words, the functionalist approach tries to explain how mental states are integrated into the physical world and how they can exert a causal influence on behaviour.

However, whatever option of "reduction" chosen, reductionist physicalism is confronted to serious difficulties regarding the states of experience, *the "qualia"*, in as much as it rejects any possible distinction between the mental and the physical states (Chalmers 1996; Levine 1983; Nagel 1974; Esfeld 2005). Regarding "identification" of mental states to physical states, the existence of the conscious, subjective experience that accompanies the brain activity to which it would be identified, if not denied (Dennett 1991), is still to be explained. And even by supposing that the states of experience could be "functionally" definable, which is far from obvious in view of our current knowledge, finding the physical realizers of a state of experience still does not explain the subjective experience that accompanies the causal work of these physical realizers. We are faced with *the problem of consciousness*, or that of *the 'explanatory gap'*, initially formulated by Levine (1983) and more recently explored by Chalmers (1996, 2007) or by Esfeld (2005, Chap. 14): we have to explain why and how can conscious experience « arise from » physical (neural) states, and why does a specific mental state (for example, pain, not itching or tickling) arise from a given brain state.

The key point is that conscious experience cannot be completely understood from the external third-person description of the brain activity provided by neurobiology. As explained by Nagel (1974) in his famous article "*What is it like to be a bat?*", a third-person description, as accurate as we want, of the bat's echo-locatory system cannot compensate our inability to experience the bat's first-person subjective auditory experience, which would require us to be bats. This inability of reductionist physicalism to account for subjective experience has also been illustrated by Jackson's knowledge argument (Jackson 1996). This argument relates the case of a brilliant neuroscientist, called Mary, who, confined in a black and white room since birth, and who has learnt, via black and white television and computers, all of the physical theories, including the wavelengths associated to colors, and the neurophysiology of vision. According to reductionist physicalism, Mary is supposed to know *everything* about perception of colors since mental states, including states of experience, can be reduced to physical states and then "explained" physically. Later, when Mary walks out of her room and sees a ripe tomato, *she has a new experience she could not have in her black and white room: she experiences the red color of this ripe tomato*. If, as rightly claimed by Jackson, Mary has learned something new by experiencing the red color of this tomato, her previous knowledge about colors and their perception was incomplete; therefore, reductionist physicalism is false.

In order to overcome the problem of consciousness it seems that we should reject reductionist physicalism, keeping the irreducibility of the mental to the physical, thus adopting a form of phenomenal dualism. However, as shown above, dualism cannot be easily reconciled with the causal interpretation of the mind-body connection: dualism of substances appeals to an elusive concept of psychophysical "interaction", while dualism of properties of a single material substance seems incompatible with the concept of mental causation. Keeping dualism of properties, in order to overcome the problem of consciousness, thus leads to a rejection of the causal interpretation on the mind-body connection. This conclusion echoes with Spinoza's neutral monism philosophy that keeps the irreducibility of the mental to the physical, conceived of as intelligible aspects of a single "neutral" substance, but does not appeal to any concept of causation for explaining their strong correlation. Along this line of thought, the mind-body connection is conceived as a *parallelism of aspects*. Spinoza's neutral monism philosophy, as well as some of its more recent reinterpretations, will be presented in the next chapter.

Chapter 2
Exploring Neutral Monism Philosophy

Substance monism asserts, in opposition with Descartes' dualism of substances, that everything in the world derives from a *single* substance and that all phenomena can thus be explained by referring to this single substance and its transformations. Asserting that this single substance is matter leads to *materialist monism*, also called "physicalism" above, while asserting that this single substance is neither matter nor mind leads to *neutral monism*.

Neutral monism philosophy has been developed very rigorously by Spinoza in the Ethics (Spinoza 1677), in the form of the deductive model of mathematical discourse, starting with a list of definitions and axioms from which are deduced propositions relating to fundamental ontological categories, such as those of substance, mode and attribute. Spinoza's neutral monism philosophy has been later re-interpreted in purely epistemic terms by Russell, by James and by Mach on slightly different grounds. Jung and Pauli have explored neutral monism from the point of view of the depth psychoanalysis (theory of archetypes) and, more recently, this philosophy has been revisited within the framework of quantum theory by Bohm and Hiley and by Atmanspacher and Primas. Let us briefly present this philosophical system, which constitutes a general framework for a consistent interpretation of the mind-body connection.

2.1 An Enlightening Illustration of Neutral Monism

Before presenting Spinoza's neutral monism philosophy and its modern reformulations and refinements, let us provide a general, illustrated presentation that will help the reader grasp some of its essential ideas. In this landscape photo we see in the distance a bird flying over a lake while two TV cameras zoom on its movements from two different angles and transmit the information to two separate monitors that you are watching in another place (Fig. 2.1).

P. Uzan, *Mind-Body Entanglement*, SpringerBriefs in Philosophy,
https://doi.org/10.1007/978-3-030-90173-8_2

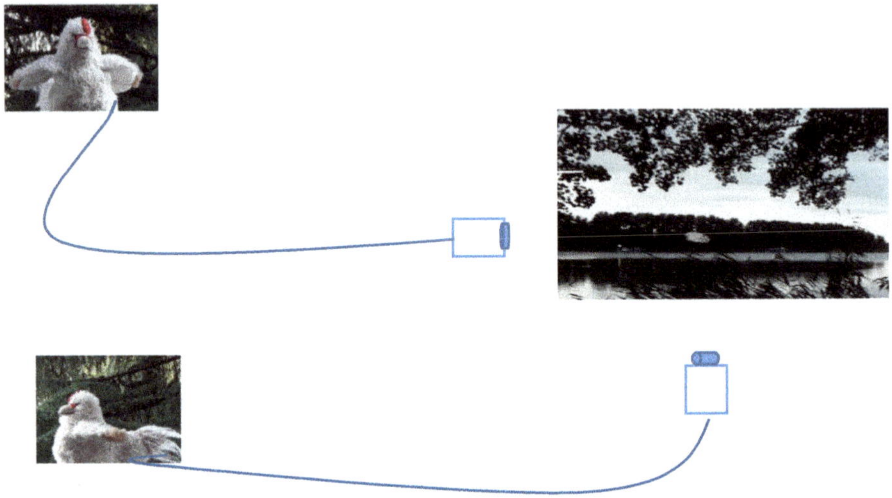

Fig. 2.1 A unique bird observed under two different aspects

If you are not aware of the real situation, it may seem that you are watching two different birds which instantaneously communicate. For example, when one of them is turning, the other makes a corresponding turn and when one of them faces the front, the other one faces the side. The explanation of such correlations is of course that *there is only one bird* observed under two different angles, giving rise to two perfectly synchronized complementary descriptions.[1] With regard to our purpose, this situation suggests that the individual's psychic and somatic features could be seen as *correlated complementary descriptions of a same underlying level of reality*. This means that their distinction can be understood as *an epistemic distinction*, relying on our means of investigation. This illustration introduces the presentation of the philosophy of neutral monism in which our book's thesis is rooted.

2.2 Spinoza's Psychophysical Parallelism

According to Spinoza, there exists only one infinite, eternal, Substance (also called God or Nature) which is neither mind nor matter. Spinoza defines the Substance as (Definition III):

> that which exists in itself and is understood by itself, that is to say that of which the idea does not need the idea of anything else, from which it would be formed.

[1] Note that Fig. 2.1 showing a bird flying from two different perspectives illustrates only one aspect of the concept of complementarity (that of complementary descriptions, which is mentioned here for introducing to Spinoza's neutral monism philosophy) but this concept also applies to the quantities that can be measured, or observables, as will be explained in detail in Sect. 4.2.

Spinoza's substance is thus conceived as not caused by something else but *causa sui*. Moreover, Spinoza's Substance cannot be born or die, it cannot be divided and limited and in contrast with Descartes' dualism of substances, mind and matter are understood as two infinite *attributes* of this single substance, that is, as two intelligible aspects of it we can perceive.

Everything in reality is thus considered as a *mode* of this neutral substance, that is, a way of being, a state or an affection of this substance. It consequently seems that for Spinoza experience plays a fundamental role since it restraints our knowledge of the Substance to only two attributes and that, moreover, the very notion of attribute relies on what we can perceive or not, which also relies on experience. Spinoza conceives of the difference between the two mental and physical attributes of the Substance as irreducible, meaning that one of them cannot be derived from the other. In addition, he asserts that these two attributes of the substance cannot be causally related:

> The body cannot determine the mind to thinking, and the mind cannot determine the body to motion, to rest, or to anything else. (*Ethics*, part III, prop. 2).

In order to justify this proposition, Spinoza refers to the irreducible distinction between the mental and the physical attributes of the Substance. He says that all modes of thought have God for their cause, insofar as he is a thinking thing, and not insofar as he is explained by another attribute (by Proposition 6, part 2), and that, consequently, what determines the mind to thinking is a mode of thinking, not a mode of extension. The body cannot thus determine the soul to thought. Similarly, the movement and the rest of the body must come from another body which itself is determined by another body with movement and rest. Consequently, everything that occurs in a body must have come from God, as affected by a certain mode of extension, and not by a certain mode of thinking. Hence the above proposition about the absence of any causal connection between the mental and the physical aspects or modes of the neutral Substance.

If Spinoza rejects the idea of a causal connection between mind and matter, he claims that the mind and the material world are nevertheless structured by causal chains. As long as things are considered as spiritual phenomena, the order of Nature is a relation of cause and effect by the attribute of the mind alone; and as long as we consider them as material phenomena, the whole order of nature must be explained by the attribute of extension alone. Insofar as there is only one single Substance apprehended according to two different and irreducible attributes, the two causal chains structuring its two distinct and irreducible modes *are necessarily connected*. Along this line of thought, Spinoza develops in the Ethics the idea of a *psychophysical parallelism* regarding the two, mental and physical, attributes of the single substance. The concept of psychophysical "parallelism" thus contains two main ideas that have been illustrated by the Fig. 2.2 of Sect. 2.1: (1) there is no causal connection between the two distinct and irreducible mental and physical attributes of the substance, whereas (2) there is a strong synchronization between the mental events and the physical events, expressed by Spinoza as follows (Ethics, part II, prop. 7):

> The order and connection of ideas are the same as those of things.

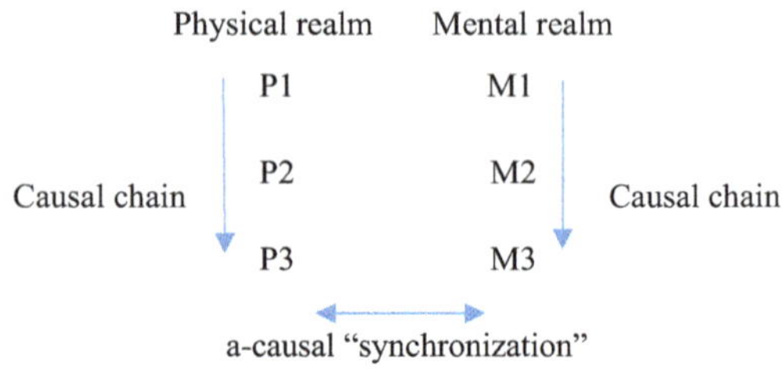

Fig. 2.2 Psychophysical parallelism

The latter psychophysical "synchronization" means that the mental and physical (causal) chains of events are coordinated or strongly correlated. In the scholium of this proposition, Spinoza reminds the reader that, in contrast with Descartes' dualism of substances, the thinking substance and the extended substance are *one and the same substance*, which is designed sometimes under one of its attributes and sometimes under the other. Consequently, whether we conceive of Nature under the attribute of extension or under that of thought, or under such other attribute as it may be, we will always find one and the same order, one and the same connection of causes of effects. Note that this idea of coordination is different from Leibniz's idea of pre-established harmony, asserting that such a coordination has been established from the beginning by God. It is also different from Malebranche's idea of occasionalism according to which this coordination would take place at the very moment these events occur, thanks to the intervention of God. In contrast, Spinoza's idea of synchronization of mental and bodily events relies on the idea that they are nothing but two aspects of a same and single Substance. They are to be understood as *two distinct but yet inseparable aspects of it*, hence their synchronization –like that of the two aspects of the single bird in Fig. 2.2.

However, in spite of this very enlightening suggestion, Spinoza does not provide more clarification on the nature of this "synchronization", and already on the meaning of this psychophysical "parallelism", apart from its essential a-causal character. Furthermore, Spinoza's psychophysical parallelism leads to the idea that any piece of matter should also have a mental state, an idea difficult to defend (Esfeld 2005, pp. 34–35). For our purpose, it will not be necessary here to discuss the objection of panpsychism that can be made to Spinoza's psychophysical parallelism. We will only focus on the question of the nature of the correlations between the mental and the physical states of a living being (a human, in particular) and not of any matter, and thus concentrate on other difficulties, including those mentioned by Jonas (1982, 1986) and Esfeld (2005, chap. 3). These difficulties can be synthesized as follows: (1) this approach gives rise to the idea of an intramental determinism, since Spinoza's idea of a strict psychophysical isomorphism and that of a physical determinism (at least, if we refer to classical physics) lead to the denial of any spontaneity of mind and of free will; (2) it could even give rise to the idea of subordination of the mental to the physical if intramental determinism is unilaterally "imported" from the physical domain, which contradicts the parity between the mental and the physical domains implicitly involved in the very idea of parallelism; (3) it thus leads to epiphenomenalism, conceiving of mind as a mere impotent epiphenomenon of matter (Jonas 1982, p. 241). In addition to these difficulties, let us note (4) the incompleteness of Spinoza's psychophysical parallelism regarding the *specificity* of the mind-body connection. To say that God (or Nature) is at the origin of these correlations does not explain why, for example, fear is associated with an increase in blood pressure and adrenalin concentration in the blood, or that joy is associated with an activation of the parasympathetic nervous system.

Is it possible to work out a model of the relation between the mental and the bodily features of the individual that overcomes the mentioned difficulties of Spinoza's proposal and can, furthermore, provide an explanation of the specificity of the observed psychophysical correlations? Let us make some suggestions that will enable us to explore this line of thought further.

The first difficulty of Spinoza's proposal previously mentioned could be overcome by asserting that the mental domain is *not* governed by laws isomorphic to the physical laws, and especially that there is *no mental determinism*. Actually, as noted by psychology, the mental domain seems to be governed by "laws" that cannot merely be put in strict correspondence with those of classical physics. These laws, like those of condensation or displacement of ideas and images that are involved in dreams (Freud 1915), do not seem to be deterministic and room for imagination and free-will is certainly left –which obviously weakens the original idea of "parallelism" that supposes a strict isomorphism of mental and bodily events. In contrast with determinism of classical physics, *indeterminism* seems to govern the mental domain. Mental indeterminism, which closely relies on the complementarity of mental observables, means that their values are not totally determined prior to their observation, that these values are revealed by an appropriate questioning performed *in a given context* (Aerts et al. 2011; Busemeyer and Bruza 2012). This indeterminism gives rise to order effects and interference effects that can be easily observed. For example, asking a subject the question A about the intensity of their sadness and then a question B about the intensity of their happiness does not provide the same results than asking these questions in the reverse order (B and then A). The complementarity of mental observables and that between somatic observables will be explained in detail in Chap. 4 where they will find a precise, mathematical expression. As will be explained in this book (Chaps. 3 and 4), the possibility of defining complementary observables in the mental and in the somatic domains grounds the property of *non-locality* of the psychophysical correlations. As a consequence of this more realistic approach to the laws that govern the mental domain, we cannot say any more that the mental is "subordinated" to the physical features of the individual (second difficulty), and in particular to its strict determinism, if any, since these physical features governed by physical laws are *not* imported or "embedded" into the mental domain, which is governed by its specific laws. Accordingly, the mental cannot be conceived as a "mere impotent epiphenomenon of matter" (third difficulty) since the mental properties are not supposed to be "caused by" or "derived from" the physical properties, as secondary properties without any role. Here, both mental and physical properties have the same status: they are distinct and autonomous aspects of the same psychophysical unity, each of them being governed by specific laws.

Turning now to the fourth difficulty, a more complete explanation of the specificity of the psychophysical correlations can be provided by adopting *an evolutionist approach*. According to this evolutionist approach[2] the strong psychophysical correlations that define the specific psychosomatic structure of any individual would have been established *during the evolution process*: in the phylogenetic evolution, as a member of the mankind; in the ontogenetic evolution, as the parent's offspring, and even in the socio-cultural evolution, through personal experience in society. As explained by Darwin (1872), and more recently by neuroscientist Lestienne (2008) and physician Janssen (2006) by referring to contemporary studies in the biology of evolution, the building of such psychophysical correlations is essential to prepare the organism in facing primordial emotions (like hunger, thirst or fear). In evolution, emotional experiences have been correlated to biological programs capable of responding to or coping with them, meaning that *the specificity of the psychophysical correlations has an evolutionary significance*. For example, the increase of adrenalin and cortisol production by the surrenal glands and a greater stimulation of the sympathetic nervous system prepares the individual to a "fight-or-flight" response to the fear triggered by a dangerous situation (Lestienne 2008).[3]

Consequently, the previously mentioned difficulties due to the rigidity of a too strict psychophysical parallelism can been resolved in agreement with the observations of psychology regarding psychic dynamics, while the specificity of the psychophysical correlations can be explained by referring to the process of evolution. Spinoza's psychophysical parallelism thus provides an appropriate framework wherein the mind-body connection can be consistently studied, without appealing to the idea of a causal "interaction" between the mental and physical features of the individual. Can concrete situations involving human's experience be fruitfully analysed within this abstract framework? Two reformulations of neutral monism philosophy respectively referring to individual and collective levels of human experience have been developed. Let us present these approaches.

2.3 Neutral Monism and Individual Experience

For Gustav Fechner, German philosopher and psychologist (1801–1897), the mental and the physical are two different but interdependent *modalities of experience* that are not necessarily linked by a causal relationship. They are considered as phenomena relating to the internal and external points of view that one would have on the same thing, as asserted in his treatise on metaphysics *Zend-Avesta* (Fechner 1851; 2: p. 135):

[2] The evolutionist approach will be developed in more detail in Sect. 3.1 from Darwin's theory of emotions.

[3] Furthermore, as will be noticed in Chap. 5, the evolutionary meaning of psychophysical correlations could explain the genesis and development of internal diseases.

> The body and the mind, or the material and the spiritual, or the physical and the psychic (all these opposites are here used in an equivalent way while retaining the most general sense) are not different according to their foundation or their essence, but only because of the different point of view according to which they are conceived or observed. What appears to oneself - from an internal point of view - as spiritual or psychical, can appear to someone who stands in front, placed from an outside point of view - only in another form, which is precisely the material and bodily expression of the same thing.

Fechner's fundamental contribution to neutral monism lies in his proposal for a psychophysics, an experimental science whose object is to discover the laws governing the connection of the external and internal modalities of experience, that is, the connection between the physiological stimuli and the corresponding psychic phenomena, the "sensations" (Fechner 1860). Fechner's psychophysics is based on the idea that psychological phenomena can be quantified just as much as physical phenomena, against the mainstream of the nineteenth century that mental phenomena could not be the subject of an experimental study. Fechner's psychophysics takes into account Herbart's discovery that sensations vary in intensity and that there is a threshold of excitation (stimulus) below which no sensation is felt. Fechner then shows the existence of a systematic relation between, on the one hand, the intensity of the physical stimulus and, on the other, the subjective sensation. In fact, to measure the latter "subjective sensation", Fechner refers to the measurement of a physical quantity (the intensity of the stimulus) which is supposed to be directly related to the change of sensation. He uses the notion of "the smallest increase of the intensity of the stimulus capable of causing a change of sensation" proposed by Weber (1834), this quantity then serving as a unit of measure for the magnitude "sensation", which is of mental order. Weber-Fechner's law states that the perceived sensation varies in proportion to the logarithm of the intensity of excitation (Fechner 1860).

Ernst Mach's interpretation of neutral monism philosophy is based on the idea that the ultimate constituents of reality, its "elements", are neither physical nor psychic (or mental) but are constituted of *the primary data of perception* (Mach 1886). For Mach, "the world consists only of our sensations" whose synthesis in "complexes of sensations" is carried out by the subject of perception. There are three types of sensory complexes: those defining physical reality (complex of colours, sounds, etc.), those defining our body and those constituting consciousness. The complexes of sensations can in fact be considered as psychic or as physical according to the way in which we determine them, this division of reality being considered as functional and not essential. In contrast with Fechner's psychophysical parallelism that considers the physical and the psychic domains as aspects of a more fundamental level of reality, Mach upholds a kind of continuity between them by asserting that:

> ...it can easily happen that the physicist who does not study the psychology of his operations ... overlooks the sensory element at the foundation of his work ... Psychological analysis has taught us that this is not surprising, since the physicist is always operating with sensations... A physical concept does not mean anything else than a certain type of composition between sensory elements... the physicist always operates with sensations. (Mach 1886).

For his part, the psychologist and philosopher William James, whose theory of emotions will be discussed in detail in Chap. 3, suggests that mind and matter are only two distinct ways of considering the same "fabric" made by "the experiment in general". Like in Mach's approach, the distinction between subject and object, thing and thought being is understood as being purely functional and non-ontological. This means that matter as much as consciousness does not exist independently of the subject who experiences the world. If, for James, material existence is ultimately reduced to what is perceived, in agreement with Berkeley's analyses, consciousness only refers to the "susceptibility of the parts of experience to be reported or known".

Russell's neutral monism philosophy, for example expressed in the following sentence drawn from his *Analysis of Mind*:

> The stuff of which the world of our experience is composed is, in my belief, neither mind nor matter, but something more primitive than either. (Russell 1921, lecture I).

also conceives of the mind-matter distinction as an *epistemic distinction* (Russell 1948, p. 209):

> My own belief is that the mental and the physical are not so disparate as is generally thought. I should define mental occurrence as one which someone knows otherwise than by inference; the distinction between mental and the physical therefore belongs to the theory of knowledge and not to metaphysics.

Standing out from Mach and James analyses, Russell proposes in *Analysis of Mind* (1921) to refer this distinction to that between the supposed causal laws by which we describe these two fields (lesson V): on the one hand, physics is appropriate to apprehend the physical world and, on the other, psychology is appropriate to apprehend the mental world. However, Russell remarks that the distinction between the laws governing, respectively, the physical and mental domains is far from clear. It may be that their distinction is only a question of degree, as it is possible, at first, to distinguish the laws governing the gases and those governing the solids whereas these two theories can be derived from one and the same theory (statistical physics). Moreover, this distinction does not stand up to the fact that an event may be the effect of several causes according to several different nested causal laws, mental or physical, and that, therefore, it seems difficult to clearly distinguish between purely physical laws and purely psychological laws.

Faced with the difficulty of establishing a clear and sharp difference between physical laws and psychological laws, Russell, always assuming the existence of causal laws *in* reality, then suggests that:

> .... mind and matter are built from a common reality whose causal laws ... form the basis from which physics and psychology are developed. (Russell 1921).

In other words, the distinction between physics and psychology would not reflect that of their object, the matter for physics and the mind for psychology, but it would be purely *epistemic*, due to the fact that we are apprehending differently, by different "logical constructions", the aspects of this neutral reality. In order to distinguish the two types of relations defining respectively physics and psychology, Russell introduces the notions of "active place" and "passive place" (Lecture VII). The sensation associated with the vision of a physical object by a subject indeed involves two elements: the place where this object is located (called "active" place) and the place where the subject is located (called "passive" place), and the "perception" of this object is its appearance from the point of view of the subject. Russell thus claims that if physics governs the relations between the different appearances of the same object, relating to the same "active" place, psychology is concerned with linking different appearances from the same point of view or from the same perspective, the "passive" place (Lecture XV).

However, if Russell's consideration of the "active" and "passive" places of, respectively, the object and the subject of knowledge refers to individual experience, it does not deepen the nature of their relationship. Such a deepening would require establishing precise correspondences between the properties of these "active" and "passive" places. Fechner's psychophysics took a decisive step in trying to relate objective stimuli and subjective sensations. His work shows that the *experimental* investigation of the mind-body connection, which involves precise mental and somatic features of the human being, is essential for such a deepening. Along this direction of research, Sect. 4.4 will present such an experimental study carried out according to current protocols. Let us now explore Jung's and Pauli's approach to neutral monism that refers to the *collective* level of experience.

2.4 Jung and Pauli: Unus Mundus and Archetypes

Swiss psychiatrist-psychoanalyst Carl Gustav Jung, founder of depth psychology, suggested that the notion of unconscious could apply beyond the personal domain. The psychoanalytic unconscious would have a collective component consisting of archaic residues of "biological development, prehistoric and unconscious, of the spirit in the archaic human, whose psyche was close to that of the animal" (Jung 1964, p. 67). It is these archaic residues of the psyche that Jung first calls archetypes, conceived as empty preforms that structure and energize the whole psychic, perceptive, affective, cognitive and behavioural processes. Archetypes are inaccessible to consciousness but acquire conscious content at certain times of life, for example during certain dreams,[4] in collective representations and in the appearance of cultural forms.

The initial concept of archetypes as "primordial psychic images" was extended by Jung in close collaboration with physicist Wolfgang Pauli. These researchers have proposed that archetypes constitute the underlying structures of the psyche *as much as those of the material world.* As explained in his book *Mysterium Conjunctionis* (Jung 1963, p. 767), the archetypes would form a neutral, psychophysical level, named by Jung *Unus Mundus,* "the one world", after Middle Ages' philosophers:

[4] For Jung, the notion of personal unconscious proves to be insufficient for interpretation of dreams.

> Undoubtedly the idea of the *Unus Mundus* is founded on the assumption that the multiplicity of the empirical world rests on an underlying unity, and not two or more fundamentally different worlds exist side by side or are mingled with one another. Rather, everything divided and different belongs to one and the same world, which is not the world of sense but a postulate…

From this primordial reality psychic events emerge as much as those falling in the material domain. According to Jung and Pauli, the psychic and material levels would be correlated by archetypes, conceived of as ordering and regulating factors constituting the most fundamental level of reality:

> The ordering and regulating factors must be placed beyond the distinction of "physical" and "psychic".[5]

In a letter to Pauli, Jung has suggested that number could be regarded as the most primitive archetypes of order, relating both material and psychic domains, making then possible the study of their connection:

> These [the natural numbers] seem to be the simplest and most elementary of all archetypes. That they are archetypes emerges from the psychological fact that simple whole numbers, given the chance, amplify themselves immediately and freely through mythological statements…[6] (Jung, 1963).

The idea that number is not only an essential aspect of the material realm but also of the dynamical aspect of psyche has been explored by Marie Louise von Franz, a close collaborator and continuator of his research, in her book *Number and Time* (von Franz 1974). For his part, Pauli developed the idea of a future science whose object is neither the material domain nor the psychic domain but the *neutral psychophysical reality*, which would then require the development of a unitary, "symbolic" psychophysical language. Pauli explains this in a letter to Rosenfeld (Pauli 1952):

> Regarding the invisible reality, of which quantum physics as well as psychology of the depths provide us with some indications, a psychophysical, symbolic and unitary language should ultimately be adequate…

In addition, Jung has suggested that the archetypes, the coordinating principles of the psychic and physical experiences, could be manifested directly by *significant coincidences between events in these two areas*. In contrast with the concept of causation that describes the succession of events, the concept of synchronicity deals with their significant coincidence, as explained by Jung in the preface of the English edition of Yi King:

> Just as causality describes the sequence of events, so, for the Chinese mind, synchronicity deals with the coincidence of events. The causal point of view tells us a dramatic story about the way D came into existence - it originated from C, which existed before D, and C in turn had a father, B, and so on. For its part, the synchronistic vision strives to produce a coincidence table that is also full of meaning. How is it that A', B′, C ', D', etc. all appear at the same time and place? This happens first of all because the physical events A 'and B' are of

[5] This sentence is drawn from Pauli's letter to Fierz, January 7, 1948, in von Meyenn (1993), p. 496–497.

[6] Translation by Meier (2001, p. 127).

> the same quality as the psychic events C 'and D', and then because all are the exponents of a single momentary situation. (Jung 1951).

As an example of synchronistic coincidence, Jung reports the sudden irruption of a beetle at his window, an extremely rare event, while one of his patients was telling him a dream in which she was given a golden beetle as a gift (Jung 1963). These "synchronistic" coincidences would be the witnesses of their intertwinement, within a same archetype or archetypal structure, in the underlying psychophysical reality, *Unus Mundus*. The concept of synchronicity thus refers to the meaningful coincidence of events arriving at once. It seems to also underlie the very foundation of Chinese theory of Yi Jing, which incorporates the essential ideas of Taoism and Confucianism. Yi-King, or the "Book of Changes", asserts that the events that tend to occur at the same time (and that the act of divination[7] makes it possible to reveal) have the same meaning. More generally, as noted by von Franz, Traditional Chinese philosophy is less concerned with establishing causal links between events than to ask "what are the things that tend to happen together in time?"[8]

2.5 Quantum-Like Neutral Monism Philosophy

The two following proposals aim reformulate neutral monism philosophy within a quantum-like framework. The underlying idea is that the key-concepts of quantum *theory*, namely those of complementarity and entanglement, can be used to clarify the basic statements of this philosophy.

2.5.1 Bohm and Hiley: The Implicate Order Theory

A first quantum-like version of neutral monism philosophy has been proposed by the two physicists David Bohm and Basil Hiley in the implicate order theory, presented in their book *The Undivided Universe* (Bohm and Hiley 1993). In their "implicate order" theory, they consider mind and matter as two manifestations or projections of the same underlying reality, their connection being interpreted similarly to the phenomenon of quantum non-locality. To understand the ideas developed in this theory we will proceed in two steps, following Bohm's presentation: the pilot-wave theory, aiming to explain the physical phenomena of the quantum domain, and then the implicate order theory, presented as its generalization and in which the mind-matter connection can be fruitfully studied.

---

[7] By manipulating yarrow stalks or, more simply, three pieces of currency in order to get six numbers allowing the construction of an hexagram, which is loaded with meaning.

[8] See M. L. von Franz (1980): *On Divination and Synchronicity, the Psychology of Meaningful Chance*.

It is, in the first instance, the desire to account for the existence of quantum correlations between spatially separated events[9] that led these authors to develop the pilot-wave theory, initially proposed by Louis de Broglie (Bohm 1980). This theory suggests to preserve the ontology of the particles undermined by the Copenhagen interpretation by postulating that the particles are endowed at any time with a well-defined position and a well-defined momentum. This means that they are treated as corpuscles whose position and momentum are simultaneously defined at each moment. In this theory, the wave function plays two roles. On the one hand, as in the standard quantum theory, it continues to play its predictive role via Born rule[10] - except that its squared module is here interpreted as a probability of *presence* of particles (not only as a probability to observe it). On the other hand, the wave function is also interpreted as a "guiding wave" or an "information field" since it is used to define, via the Schrödinger equation governing the temporal evolution of states, a *quantum potential* that contains all the information on the experimental situation necessary for organizing or piloting the quantum phenomena through "forces" (Bohm and Hiley 1993, §3.2). The quantum potential would be responsible for the typically quantum effects, such as the EPR correlations in Aspect's experiment (Aspect et al. 1982) or the interference phenomena. For example, in Young's double slits experiment the interference phenomenon can be established from the expression of the quantum potential defined for this device (Philippidis et al. 1979). In this model, the particles (which are classical corpuscles) follow continuous trajectories that are affected by the quantum potential encoding all information about the experimental setup, information called "active information"[11] by Bohm and Hiley (1993, §3.2). Active information thus plays the role of in-forming or shaping quantum phenomena from an underlying level of reality of a purely informational nature.

Bohm and Hiley generalized the idea of existence of an underlying level of reality that would contain all the information needed to guide the spatio-temporal quantum phenomena. They generalized the notion of "quantum potential" defined within the strict quantum domain to that of "implicate order", a deeper order of reality that would contain all the information necessary to guide the emergence of *our manifest world.*[12] That is, the phenomenal world, which is a visible, tangible order level of reality is regarded as "unfolded" from the implicate order level. In particular, all the events of our manifested world, *whether material or mental*, are considered as projections of an underlying undivided reality that would be "enfolded" in the implicate order.

---

[9] Two events are said to be "spatially separated" if they cannot have any causal connection according to relativity theory.

[10] Born rule allows the calculation of the probabilities associated with the possible outcomes of a measurement.

[11] The idea underlying the notion of "active information" is that a small amount of information and energy can produce much larger effects.

[12] This means that the quantum potential can be considered, *for the purely physical domain*, as the mathematical expression of the implicate order.

The dynamical process of enfoldment-unfoldment of the mental and material aspects of reality into the implicate order has been very well illustrated by the hologram metaphor (Bohm and Hiley 1993). A hologram is a three-dimensional photograph of an object. To make such a photograph, the interference pattern of two light beams which, respectively, arrive directly and after reflection on the object is recorded on the photographic plate (Fig. 2.3). When this interference pattern is illuminated with laser light, it produces a three-dimensional image of the original object (Fig. 2.4).

Fig. 2.3 A hologram

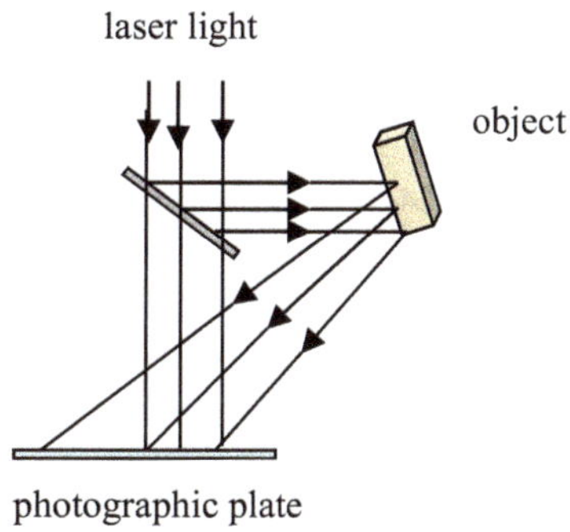

Fig. 2.4 A hologram: 3-D image of the object

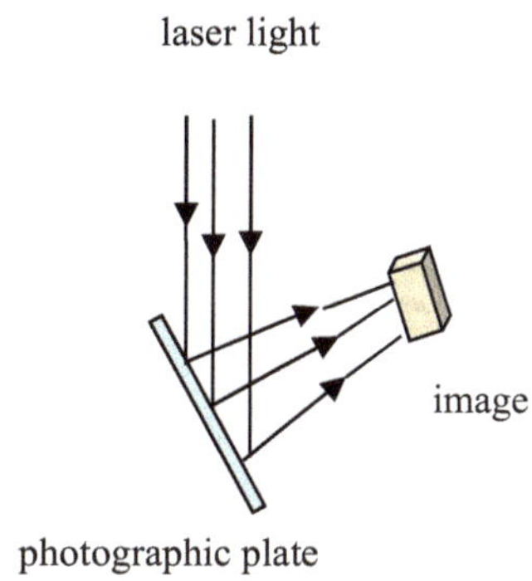

The two pieces of information respectively carried by the two light beams are encoded in the seemingly meaningless entangled swirls of the interference pattern. Illuminating it with laser light allows making the appearance of the three-dimensional image of the object (Fig. 2.4). Similarly, one can say that the psychical and the physical aspects of the individual are enfolded in an inaccessible, underlying, neutral level of reality from which they co-emerge and appear as distinct by appropriate means of investigations. According to Bohm and Hiley, this universal process of unfoldment-enfoldment, called "*holomovement*", would be the heart of reality. In the holomovement, all aspects of our manifested world would emerge from a deeper order of reality and end up re-enfolding in it, their apparent stability and autonomy being one of the consequences of this incessant movement. Bohm and Hiley illustrate the notion of holomovement in the following way:

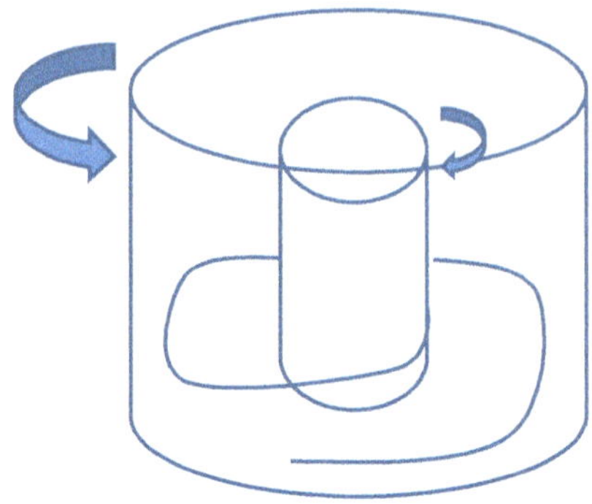

Fig. 2.5 The ink spot model, an illustration of Bohm's and Hiley's notion of holomovement

The annular volume between two concentric cylinders is filled with a very viscous liquid such as glycerin (Fig. 2.5). Suppose the outer cylinder can be rotated while the inner cylinder is fixed. An ink droplet is placed in the fluid and the outer cylinder is rotated slowly. The ink droplet will then be stretched to a finer and finer shape until it disappears. However, if the inner cylinder is now slowly rotated in the opposite direction, all the steps in the process of disappearing the ink droplet will reappear in the opposite direction, from the finest form to its original form (the quasi-reversibility of this phenomenon being possible by the fact that the ink diffuses only very little in the glycerine). In this experiment, the manifest, explicit order of the ink droplet is not destroyed when it disappears from our gaze but it is enfolded in glycerin in a hidden order.

This experiment can be done with a series of droplets that are enfolded consecutively and regularly, after the same number of turns, during the rotation of the outer cylinder. When its direction of rotation is reversed, the droplets reappear one after the other; but if the reverse rotation is carried out sufficiently rapidly, a stationary drop of ink appears for a certain time, giving the appearance of a stable "particle". In addition, if the successive droplets are now placed adjacent to the glycerin, the reverse rotation of the inner cylinder shows a particle moving along a continuous path. This continuous displacement is in fact an illusion generated by the fast and regular sequence of unfoldment-enfoldment processes. It is possible to specify the dynamics of this experiment by defining a degree of enfoldment of each droplet by a parameter measuring the number of revolutions of the inner cylinder necessary for the droplet to reappear in its original form (Bohm 1980, p. 162).

The holomovement does not only explain the emergence and apparent stability and autonomy of the physical world from a more subtle level of reality. It is also applied to *consciousness* with "… its constant flow of evanescent thoughts, its feelings, desires and impulses" (Bohm and Hiley 1993, § 15.12). As with physical processes, perceptions, thoughts, and all mental processes are connected by a relation of unfoldment-enfoldment: each of them combines and separates from others, and in a sense contains all others -as the simultaneous presence and the activity of the traces ("reverberations") left by the successive notes of a piece of music in consciousness produces the sensation of a continuous movement and a general sense.[13] Active information, initially defined in the more restricted framework of the pilot wave theory (see above), would then guide all the mental and physical aspects of the individual.

The interweaving of physical and mental processes within the same individual leads these authors to answer the question of their relationship as follows (p. 386):

> The answer we propose is that there are not two processes. Rather, it is being suggested that both are essentially the same. This means that that which we experience as mind, in its movement through various levels of subtlety, will, in a natural way ultimately move the body by reaching the level of the quantum potential and of the "dance" of particles. There is no unbridgeable gap or barrier between any of these levels...Thus, there is no division between mind and matter, psyche and soma. The common term psychosomatic is in this way seen as misleading, as it suggests the Cartesian notion of two distinct substances in some kind of interaction.

For the human being, all this implies a deep unity in which the mental and physical aspects participate in each other. The intellect, the emotions and the overall state of the body are considered to partake and to take part[14] in a same fundamental flow, so that there is no real division between psyche and soma. Bohm and Hiley represent this idea of interpenetration of physical and mental aspects of reality through *the universal holomovement process,* putting thus a strong emphasis on the concept of movement, along the line of process philosophy developed from Heraclitus to Whitehead (Whitehead 1929).

2.5.2 *Atmanspacher and Primas: Symmetry Breaking and Co-Emergence of the Psychic and Physical Aspects of the Psychophysical Unity*

Let us also mention another quantum-like proposal for modeling the co-emergence of the complementary physical and psychic aspects of reality from a neutral psychophysical level. This proposal has been presented by Harald Atmanspacher and Hans Primas in their respective articles entitled "Mind and Matter as Asymptotically Disjointed, Inequivalent Representations with Broken Time-reversal Symmetry" (Atmanspacher 2003) and "Time-entanglement between mind and matter" (Primas 2003). These authors suggest modeling this process by evoking a symmetry breaking mechanism of the primordial reality level, called *Unus Mundus* after Jung's and

[13] Note that in as much as consciousness is conceived of by Bohm and Hiley as a manifestation of the universal process, the holomovement, these authors logically suggest a form of panpsychism: even the elementary particles would have a "mind" that develops and strengthens when subtler levels of manifestation are attained (page 386).

[14] The verb used by Bohm and Hiley to translate this idea of interpenetration between mental and physical aspects of reality is "participate" which they recall in this same page 386 the two meanings of "partake of" and "take part in".

Pauli's proposal presented above. The mental domain and the material domain would co-emerge from this indivisible level of reality and their correlations, interpreted in terms of entanglement relations, could be understood as remnant of their common origin.

The great interest of this approach formulated within the framework of quantum theory is the explicit use of the closely related concepts of complementarity and entanglement for dealing with the mind-matter connection. To clarify this point, Atmanspacher mentions the proof of the generalized version of Bell theorem carried out by Landau (1987) - and commented by Primas- which closely links these two concepts. This theorem establishes that the non-commutativity of two pairs of observables respectively defined on the two parts of a system is a sufficient condition of the existence of non-local correlations (of EPR type) between these observables. The demonstration of this result is carried out within the framework of an algebraic representation which, although abstracted from classical mechanics and quantum mechanics (C* algebra of observables), no longer refers to the "physicality" of the systems considered. This demonstration neither appeals to a physical quantity, like energy of momentum, nor to the Planck constant that delineates the classical and the quantum domains. It only uses the *mathematical formalism* of quantum theory that represents the observables we can measure as operators of a C*-algebra.[15] Such a very general demonstration thus establishes that (Atmanspacher 2003):

> If [a system] X is decomposed into two subsystems I and II with, for each of them, non-commutative properties, then there are non-local correlations between the two subsystems I and II.

If, as proposed by Primas, the condition that the subsystems considered are of a physical nature is abandoned, it becomes possible that I and II designate, respectively, the physical and psychic domains of the manifested reality that would come from X, designating then a level of reality neutral with respect to the psychophysical distinction (Primas 1996). And, insofar as there is a sense in defining non-commutative observables for the psychic domain,[16] there would then exist non-local psychophysical generic correlations.

2.5.3 *Time Entanglement of Mind and Matter?*

As emphasized by Mc Taggart (1908), two concepts of time seem to be radically opposed. Physical time parametrizes material processes and its key features, it can be derived from theoretical physics by referring to: (1) the invariance of the fundamental laws of physics by time translation (2) their invariance by time reversal (3)

---

[15]An intuitive presentation of quantum theory is provided in Chap. 4 and Appendix 1.

[16]This question will receive a positive response in Chap. 4.

the absence of becoming in the block-universe conception of relativity[17] and (4) the report of events by tenseless sentences ordering them by the relations "earlier than", "simultaneous with" and "later than" on a time axis set for eternity. These features of physical defines Mc Taggart's B-series.

In contrast, mental time is related to our inner experience of time since a sequence of mental events is defined *for a particular subject.* Augustine upheld in *Confessions* the idea that time exists only in the human mind's apprehension of reality and not in reality itself (Augustine of Hippo 397–400 AD), whereas Bergson opposed his concept of "creative duration", which directly refers to our intuitive perception of time, to that of physical time (Bergson 1922). Mental time refers to a consciousness-related "now" and seems to irreversibly flow from our "past" to our "future". Mental events are reported by tensed concepts (Mc Taggart's A-series) since they refer to *our subjective experience of time.*

A-time and B-time series are thus to be regarded as *complementary representations* of time since they exclude each other, as mentioned by philosophers, but are both required to describe the world (i.e. physical *and* mental phenomena). Each of these concepts of time has its domain of validity and, as stressed by Primas (2003), asking which of them would be the "correct" representation of time is meaningless. This author then considers that the mind-matter intertwinement can be expressed by that between A-time and B-time that respectively refer to our internal, subjective experience and to the external description of science. Actually, as explicitly shown by Primas, *A-time and B-time series can be parametrized by a same time parameter which takes part to both series* -hence the idea of *time-entanglement* between mental and material aspects of reality.[18]

---

[17] This absence of becoming is, for example, expressed by Weyl's famous sentence: "the objective world simply is, it does not happen." (Weyl 1949)

[18] Note that Primas' concept of time-entanglement, specifying the mind-matter connection using a single time parameter, strongly echoes with Jung's notion of "synchronicity", presented above, which refers to the existence of significant temporal coincidences between inner mental events and outer events (Jung 1963).

Chapter 3
Mind-Body Entanglement

The philosophy of neutral monism that has been presented in Chap. 2 with its different interpretations and reformulations seems to constitute an appropriate metaphysical framework wherein the question of the mind-body connection could be coherently and fruitfully explored. However, these developments are not very precise regarding the individual features of the human being and their specific correlations. Either they deal with the mind-matter relationship, like in Spinoza's psychophysical parallelism, or, when human experience is mentioned, they deal very generally with it. They deal with the general relationship between experienced sensation and intensity of stimulus (Fechner 1860), its "modality" (Mach 1886) or its "internal and external points of view" (Russell 1921), but without precise references to the content of individual experience. This is also the case of Jung's and Pauli's approach, which is grounded on collective human experience (Jung 1963). As to the quantum-like reformulations of neutral monism, if they introduce a very fruitful framework wherein the non-classical concepts of complementarity and entanglement characterizing the mind-body connection can be reliably expressed, they only provide very few references to the specific connection between a person's mental life and bodily processes (Bohm and Hiley 1993; Atmanspacher 2003; Primas 2003). This general framework is now made more precise by focusing on this specific interdependence that characterizes the notion of psychosomatic unity.

3.1 The Psychosomatic Unity of the Individual

Holistic approaches to the human being and to their diseases have been explored for a long time. Ancient Greek physician Hippocrates, who is considered as the father of modern medicine, conceived of diseases as a loss of global harmony of the body and the soul, which could be restored by taking into account the patient's "temperament" and personal experiences. According to Hippocrates, health is governed by

P. Uzan, *Mind-Body Entanglement*, SpringerBriefs in Philosophy,
https://doi.org/10.1007/978-3-030-90173-8_3

the balance and interworking of the four "humors" (blood, phlegm, yellow bile, black bile) that govern the metabolism of the four elements (air, water, fire, earth) in the human body–these "elements" being defined along Aristotle's classical theory (Aristotle 350 BCE). The predominant humor in a human being determines his temperament: bilious (or choleric) for yellow bile, sanguine for blood, phlegmatic for phlegm, and melancholic for black bile. The four possible temperaments (bilious, sanguine, phlegmatic and melancholic) refer as much to physiological or morphological features of a subject as to its psychological ones. They define the fundamental psychophysical types in Hippocrates' medicine. A way of life is suggested for each of them in order to maintain good health and prevent diseases.[1] For example, the sanguine temperament for a subject results from the predominance of the humor blood and corresponds to the element air; it is physically characterized by a rather massive and well-formed body and by a direct and warm gaze; while on the mental plane, it describes an optimistic, extremely sociable and lively, enthusiastic and impulsive being, whose exuberance can easily lead to anger. The mental and physical health of a sanguine subject essentially relies on its circulatory and respiratory system and is prone to develop bleeding disorders and asthma. In addition to prescribing a specific diet for preserving good health to sanguine subjects, Hippocrates advises them to learn moderation in all areas of life.

We also find origins of the holistic approach to the human being and the diseases she can contract in Eastern philosophies and medicine: Traditional Chinese Medicine and its system of acupuncture meridians, Indian Ayurvedic medicine with its chakra system, both of them referring as much to body organs as to mental dispositions (Vasant 2002). According to Ayurveda the human being is regarded as the combination of three elemental substances or "energies", called doshas, whose balance and interactions govern conjointly its somatic and emotional features. Similar to the circulation of the Chinese "energy", the qi, along the meridians, the uninterrupted circulation of the Indian prana in the network of nadi[2] would be responsible for the harmony between body and mind (Chopra and Doiphode 2002). Ancestral practices, like Yoga, Tai Chi and Qi Gong, have been developed in order to maintain such a harmony. Yoga is a 5000-year-old discipline that finds its origin in India. The word "Yoga" comes from a very ancient Sanskrit root, "jug", that means: to connect, join, unite, put together, especially the body, the heart and the spirit. Yoga can be practiced according to several forms and aims to the realization of self-fulfillment through the unification of all the aspects of the human being, whether physical, psychic or spiritual. The practice of correct breathing, associated with specific body postures and concentration, allows the opening of the chakras and the stimulation of the circulation of "vital energy". Similarly, the practice of Qi Gong and Tai-Chi,

[1] Note that besides diet and medicine, and according to the pre-classical Greek motto "A healthy mind in a healthy body", Hippocrates offered his patients to practice physical exercises to maintain good health.

[2] The "energetic" network of nadi is organized around the *chakras,* defined as the "focal points where psychic forces and bodily functions merge with and interact with each other" (Encyclopedia Britannica).

which date back to ancient times in China, are based on knowledge and mastery of breath. They associate slow movements, breathing exercises and concentration in order to achieve a state of psycho-corporal integration for maintaining good health, preventing and healing illness, and thus increasing longevity and quality of life. Both mind-body practices, like Traditional Chinese Medicine, are grounded on the theory of five elements[3] according to which postures and movements are related to specific organs and corresponding emotions.

Let us now turn to the more recent Western origins of the concept of psychosomatic unity. First of all, it should be mentioned that in contrast with the assertion of a radical separation between mind and body, regarded as two independent substances (see Chap. 1), Descartes has also mentioned their "intimate union" in his Traité des Passions (Descartes 1649) and his Principes de Philosophie (Descartes 1644). According to Descartes, if mind and body are two separate, independent substances that must be distinguished from a metaphysical point of view, their actually are unified in all human experiences of life. This intimate union cannot be understood intellectually (Descartes 1953, letter of 28 juin 1643) but only be experienced within ourselves through emotions, passions and sensations, which cannot be exclusively attributed to one of these two categories but seem to belong to both of them:

> … we also experience within ourselves certain other things which must not be referred either to the mind alone or to the body alone. These arise, as will be made clear later on, in the appropriate place, from the close and intimate union of our mind with the body. This list includes, first, appetites like hunger and thirst; secondly, the emotions or passions of the mind which do not consist of thought alone, such as the emotions of anger, joy, sadness and love; and finally, all the sensations, such as those of pain, pleasure, light, colours, sounds, smells, tastes, heat, hardness and the other tactile qualities (Descartes 1984, I).

Descartes thus appeals to a concept of "union" of mind and body for dealing with human experiences that cannot be exclusively assigned to one of the two alleged independent, "primitive" created substances, the thinking substance and the extended substance, mind and body. It seems that an important step in the development of the concept of psychosomatic unity has been crossed here. Recognizing the existence of such global, psychophysical features of the individual that cannot be exclusively assigned to one of the two mental or bodily categories leads to the essential idea that mind and body cannot really be regarded as "independent"–which indeed contradicts Descartes' initial statement.

Along this line of thought, let us mention Freud's concept of drive he introduced in his Three Essays on the Theory of Sexuality (Freud 1905). According to Freud, drive is a psychic thrust whose source is a somatic excitation generated by a precise area of the body and whose aim is to reduce this bodily excitation by mean of an

[3] The theory of Five Elements asserts that all phenomena in the universe can be analysed as the movement and mutation of five qualities: wood, fire, earth, metal, and water.

appropriate object. His Metapsychology (Freud 1915) deals with the concept of drive that involves both psychic and bodily processes. He defines this concept as[4]:

> … a limit concept between the psychic and the somatic, as the psychic representative of excitations coming from within the body and reaching the psyche, as a measure of the work requirement imposed on the psychic as a consequence of its bodily connection.

The term "psychosomatic" suggested by the German psychiatrist Heinroth aims to capture this idea of a close interdependence of the psychic life and the functioning of the body. Heinroth wanted to emphasize the necessity to take into account the role of psychism in the appearance and development of illness, in contrast with the organicist movement of nineteenth-century medicine (Heinroth 1818). More specifically, Heinroth aimed to explain the "influence" of sex drive on physical diseases, like in particular tuberculosis, epilepsy or cancer. This view has led to the extensive medical and psychoanalytic studies of the nature of the mind-body connection. An important contribution to this view has been provided by Freud through the concept of *hysterical conversion* denoting the "jump into the organic", through somatic symptoms, of some unconscious psychical excitation (Freud 1895). According to Freud, hysterical symptoms are nothing but the somatic expression of the libidinal energy, which gets rid of the original repressed sexual representations. More radically, Groddeck suggested that *any* organic disease is the expression of a psychical disorder (Groddek 1923). He put forward the idea that the physical ailments suffered by patients symbolically correspond to hidden tendencies of the "that", while pathologies would be a way to survive, thereby alluding to a symbolic language of disorders that needs to be decoded. Along this line of thought, each physiological process would thus be directly or indirectly influenced by psychical stimuli. Physician and psychoanalyst, co-founder with Flanders Dunbar of the Psychosomatic School of Chicago, Franz Alexander has also adopted a holistic approach to the individual and the diseases they can develop. He upholds the idea of an *interpenetration* of all the features, whether psychic or somatic, of the human being, which should then be regarded as an *interdependent whole.* As a consequence, every disease should be regarded as being psychosomatic in as much as its origin and its evolution fundamentally rely both on psychological *and* somatic factors (Alexander 1950, chap. VIII). Alexander suggests that such a psychophysical interpenetration finds its origin in an evolutionary identity between psychic and physiological processes, as well as in a structural link between disease and psychic organization. For example, he upholds the idea that the primary manifestations of the psyche are "printed" in the physiology of the organic functions corresponding to the satisfaction of the vital needs.

The idea according to which the psychic organization plays an essential role in the birth of organic illness and in the healing process has been explored since the sixties by Marty and other French psychoanalysts who founded the Psychosomatic School of Paris (Marty et al. 1963). On the basis of clinical observations and

[4]This definition is given in one of the four articles forming his Metapsychology (Treibe und Triebschicksale 1915).

theoretical models, these researchers advocated a holistic approach to illness grounded on the idea that the somatization process would be triggered by the absence of a sound regulation of mental excitation by the psychism. More precisely, they upheld the idea that the proneness to develop psychosomatic disorders relies on a lack of symbolization.[5] Along this line of thought, proneness to develop psychosomatic disorders has more recently been attributed to the subject's alexithymia,[6] which denotes a difficulty in the identification, description and expression of its emotions accompanied by poor phantasmatic activity (Taylor et al. 1997). Mc Dougall has defended a similar idea about the essential link between disease and default of symbolization, by comparing the somatization process with realization of act-symptoms, such as addictions or bulimia (Mc Dougall 1986):

> By default of symbolization, the psychosomatic act-symptom remains petrified in the body. Such loss of psychic vitality, then, may crystallize the forces of anti-life that doze in the depths of man.

On experimental grounds, and without exclusively focusing, like in most of the psychoanalytic approaches, on the "somatisation" process, psycho-neuro-immuno-endocrinology has established that the psychic life is strongly correlated to the whole bodily functioning, involving its physiological, metabolic, endocrine and immunological processes. Examples of this close interdependence have been provided above when we mentioned the somatic correlates of positive and negative emotions (see Introduction). Striking evidence of this radical inseparability between mental and somatic processes is provided by the so-called "placebo effect", where a pharmacologically inert substance such as a sugar tablet or a saline solution can produce the same physiological effect than that of a genuine pharmacologically active substance (such as an antibiotic). This "effect", which has been observed in rigorous, double blind experiments, essentially relies on the physician's belief in the treatment and the patient's faith in the physician, that is, on purely psychological factors (Janssen 2006, pp. 33–49). We can also mention the phenomenon of biofeedback showing that the will of a subject that visualizes the real-time plot of a physiological variable on a screen (such as that of their heart rate or their cerebral activity) may be sufficient to modify this plot and cure some organic dysfunctions (Chiarioni and Whitehead 2008; Kranitz and Lehrer 2004). In turn, as is well known, any change in the physiological processes can have an "effect" on psychism. For example, as reported by psychiatric studies, a change in the release process of neurotransmitters, which can be triggered by drugs, correlates with an alteration of the mental state (Tung and Ric 2007).

This thorough interpenetration of the psychic and the somatic features of a person can be observed on different levels. If, as daily observed, experienced emotions

---

[5] Pierre Marty and Michel de M'Uzan introduced in 1962 at the Barcelona psychoanalytic congress the term of "operational thinking" to denote this lack of symbolisation and of phantasmatic activity that some psychoanalysts have conceived as a defensive system aiming to protect the subject from the anxiety relative to a loss of identity (Mc Dougall 1989).

[6] Alexithymia means literally "without word".

are obviously correlated with physiological and neurological variables,[7] more complex psychic states and mental disorders are also strongly correlated with the biological programs governing the functioning of the body. Such complex correspondences ground some contemporary healing practices. For example, Fiammetti's somato-emotional practice of osteopathy is based on the idea that unexpressed emotions "imprint themselves" in the body and that, consequently, deciphering the "body language" could facilitate a somato-emotional rebalancing (Fiammetti 2004). Let us also mention Janov's famous "primal therapy", which releases the repressed emotions that has been memorised in his patient's bodies by exploring their experience of life between birth and childhood (Janov 2000). Last but not least, Broom (2018) has developed a "mind-body" approach to physical diseases grounded on the idea that some of them, called symbolic diseases, are to be considered as the somatic expression, the "material signifier", of lived experience and emotional suffering of the patient.[8] This semantic approach to the mind-body connection has led him to develop a therapy based on the analysis of the meanings that the subject attributes to their illness due to their life experience.

3.2 Mind-Body Interdependency: The Case of Emotions

In order to study in more detail the mind-body interdependence, we have to focus on the individual's global, psychophysical experiences whose existence grounds their psychosomatic unity. This idea extends Descartes' suggestion that the "union" of the body and the mind is grounded on the existence of such psychophysical experiences. Let us for example focus on the notion of emotion that has been studied in detail by psychology and biology.

The word "emotion" is derived from "motion", meaning movement, a term that surfaced in the thirteenth century in French and Saxon languages. It carries the idea of an accomplished movement, the Latin root *emovere* meaning "put in motion". As emphasized by Descartes, emotions sit at the mind-body interface and express themselves through both bodily changes and subjective experience. The concept of emotion refers to a wide range of psychological and somatic changes in reaction to internal or external stimuli, hence the great diversity of its definitions. The first theory of emotions is Charles Darwin's evolutionary approach (Darwin 1872) according to which emotions confer a *selective advantage* to human being. Darwin suggested that emotions survived evolution because they guarantee adaptation to the environment, thus allowing humans and animals to survive and reproduce. For example, to be afraid of predators is essential for the survival of the individual and of the species, and a feeling of love leads people to seek to couple and to reproduce.

[7] A precise study of the correlations between the intensities of emotions and the change of cardiovascular variables will be presented in Chap. 4.

[8] Broom's approach can be regarded as an implementation of Chiozza's linguistic metaphor of the mind-body connection (Chiozza 1976).

Illustrations of the essential role of emotions can also be provided in daily life: to feel pleasant or unpleasant emotions allows for a better evaluation of the external world that helps us to organize our action; to read on our colleagues' face the emotions they feel allows a better understanding of the situation, and a better choice of behaviour. From a biological point of view, neuro-physiologist Antonio Damasio has shown that the function of emotions is indeed twofold: if the first function is, as mentioned above, the production of a specific reaction to the inductive situation, the second one is the regulation of the internal state of the body so that it can be ready for the specific reaction, as for example increasing blood flow in the arteries of the legs so that the muscles receive more oxygen and glucose, in the flight reaction. Damasio thus emphasizes that at the most basic level, emotions are part of homeostatic regulation.

These evolutionary considerations on emotions and the related biological descriptions are generally accepted as fundamental data for their study. However these considerations do not address at all the question about the nature of the mind-body connection. And concerning the behaviourist approach, this question is purely and simply eliminated since an emotion is defined as:

> a hereditary pattern-reaction… involving profound changes in all physiological mechanisms, and particularly in visceral and glandular systems (Watson 1925).

We will then focus on the theories of emotions that *explicitly* deal with the question of the mind-body connection. These theories can be classified into three classes. The first one, developed by James and Lange, asserts that visceral or, more generally, somatic aspects of emotions are primary while their phenomenal or subjective aspects would be derived from them. This first class of theories obviously relies on the materialist (or physicalist) position that has been analyzed in Chap. 1. The second class of theories, developed by Canon and Bard, asserts that the phenomenal and the somatic aspects of emotions are not causally related and can occur simultaneously. The third class of theories relies on both previous classes according to different variants, in particular by emphasizing the cognitive processes involved in emotions. Let us now present these approaches in more detail, in order to deepen the question of the nature of the mind-body connection.

Physiological theories of emotions can be traced back to psychologist William James (1902–1910). James suggested that emotion is the awareness of the physical changes activated by a stimulus. James-Lange's theory of emotions, proposed independently by William James and Carl Lange (James 1895), asserts that the phenomenal aspects of emotions, like the experience of fear or anger, originate *from* physiological reactions to stimuli and thus *follow* these visceral disturbances. As we experience different life situations, our nervous system gives rise to physical reactions associated with them, like for example increased heart rate or tremors, which would be "interpreted" or consciously perceived as specific subjective feelings. Along this line of thought, let us mention Damasio's proposal of a purely physiological definition of emotions as:

> ... many changes in the chemical profile of the body, as well as changes in the viscera and degree of contraction of the various striated muscles of the face, throat, trunk and limbs. But

they also depend on the transformations undergone by all the neuronal structures at the origin of such changes and are likely to modify in return several neural circuits of the brain itself (Damasio 1999).

In agreement with James-Lang's theory, Damasio supports the view according to which emotions, generated in extra-cortical structures from the information they receive about the state of our body, would be *causally "responsible" for the constitution of the sense of self*: this influx of emotions is sent to the cortex which integrates it into all of its cognitive or motor tasks and would thus give rise to the feeling that we are the actors of these experiences.

However, physiologist Walter Canon (1927) has pointed out several objections to James-Lange's theory. He first points out that emotions can be felt without conscious perception of physiological changes since the alteration of the connections between the viscera and the central nervous system, which makes possible conscious perception, do not notably affect emotional content. He also points out that there is no obvious specificity of the organic changes that accompany different emotions since the same visceral changes can appear in both emotional and non-emotional states. For example, if fear is associated with higher heart rate and increased sweating the same physiological reactions can accompany other emotions, such as anger. This means that fear cannot specifically be regarded as the effect of these physiological changes. Along this line of thought, Canon notices that the artificial induction of visceral changes that are supposed to be specific of some strong emotions does not actually produce them. He justified this point by referring to Marãnon's work (1924) which shows that, following an injection of adrenaline, subjects report "a simple subjective perception of certain somatic disorders which give rise to an indefinite emotional sensation". Canon also points out that subjective responses may occur too quickly to simply be the products of physical states by referring to the observation that when someone perceives a danger in the environment the feeling of fear often occurs before starting to experience the physiological symptoms associated with it (Cannon 1987).

Most importantly for our purpose, Cannon-Bard's theory of emotions asserts that the seat of emotion lies in subcortical regions[9] from which parallel innervations start, triggering both physiological (peripheral) activity *and* subjective emotional experience (Cannon 1927; Bard 1934). More precisely, according to Canon, the thalamus sends *at once* signals to viscera and to the brain, thus triggering at once the emotional experience and its physiological correlates. This means that *the psychological experience of emotion and the associated physiological changes are not causally related and that they can occur simultaneously* (Friedman 2010).

Several refinements of these two theories or combinations of them that explicitly refer to the cognitive aspect *of consciousness* have also been proposed. For example, Schachter-Singer's theory of emotions (Schachter and Singer 1962) asserts that

[9] Originally, Cannon assigned this role to the thalamus and Bard to the hypothalamus, but now this role is assigned to the whole limbic system that groups all the subcortical structures involved in the generation of emotions (MacLean 1949).

emotion would result from the "interaction" between a physiological arousal and the causal attribution of this arousal to some features of the stimuli, which is of *cognitive* order. In theories of cognitive assessment emotional experience relies on personal significance assigned to the situation lived by the subject and their past experiences (Lazarus 1991). These theories do not generally refer to the physiological changes associated with the emotional experience, but explicitly refer to a notion of signification associated to life situations, a notion that plays an essential role for understanding the specificity of psychophysical correlations.

In fact, all the approaches presented above provide relevant knowledge about emotions and help to fully characterize them. The double, physiological-psychological, aspect of emotions certainly finds its roots in the evolution process (evolutionary approach)[10] and is shaped by the subject's memories and their interpretation of the situation (cognitive approach) that always relies on a specific environmental and socio-cultural context. Moreover, emotions, which are generally expressed by mimics on the subject's face, are grounded on physiological and neuro-endocrine changes that are in turn governed by specific activities of their brain (physiological approach). However, besides this consistent representation of emotions involving their complementary aspects, there is no consensus neither on the temporal order of their bodily and mental expressions nor on the supposed "causal" nature of their connection. Actually, if subjective and physiological aspects of emotions must be considered together in order to fully characterize them, the question of the nature of their connection, closely related to that of the existence of a temporal order in their appearance, does not find a consensual response. As mentioned above, James-Lange's theory of emotion asserts that *feelings are the result of physiological changes* triggered by a stimulus. This view is supported by Damasio's biological approach according to which physiological changes would be *causally responsible* for the constitution of the sense of self. However, in Cannon-Bard's theory of emotions the psychological experience of emotion and the associated physiological changes *are not causally related and can occur simultaneously.*

On experimental grounds, the least we can say is that the existence of a causal relation between physiological and subjective aspects of emotions and of their temporal order is far from being clearly established. Wegner's experiments mentioned in Chap. 1 already shows us the need to step back on the issue of mental causation. These experiments warned us against the frequent confusion between, on the one hand, the *apparent* causal connection of conscious thoughts and associated physiological changes, and on the other, their *real* connection. As a matter of fact, contradictory observations seem to justify this warning since *both temporal orders have been reported.* In some cases, it seems clear that the physiological changes caused by taking medication are *followed* by a subjective experience. For example, negative feelings associated with anxiety are diminished *after* neurological changes (in the release of neurotransmitters) induced by taking an anxiolytic, and, similarly, the

[10] The evolutionary origin of the psychophysical correlations involved in emotions will be explored in more detail below (Sect. 5.2) when discussing the concept of body memory.

physiological changes triggered by the stimulation of certain points of acupuncture have *subsequent* anti-depressive effects. However, in other cases it has been observed that purely psychological factors can be *followed* by real physiological changes. It is for example the case in the placebo "effect" where the patient's trust in the physician and in the prescribed drug can have spectacular physiological effects, such as eradicating pain or curing a duodenal ulcer.[11] It has also been observed that a psychological stress gives rise to a *subsequent* overdrive of the sympathetic nervous system, which triggers an increasing of adrenalin and cortisol production that have measurable, generally harmful, effects on the body (Selye 1978). Even more, a simple memory or the anticipation of a future difficult situation, which generates negative emotions, can be followed by important physiological changes, such as physical symptoms of anxiety including increasing heart rate and blood pressure, sweaty palms, dry mouth, and shaky hands. Subsequently, as suggested by Cannon-Bard's theory and in agreement with the logico-philosophical analysis of Chap. 1, it seems that *no causal order* between the subjective expression of an emotion and the associated bodily changes can be definitely and universally asserted. Generally speaking, the thorough interdependence of the psychic and the somatic features of any lived experience *cannot be consistently interpreted as a direct causal connection* -a conclusion that will be checked experimentally (see Chap. 4). A new a-causal interpretation of the mind-body connection that explicitly accounts for the occurrence of *psychophysical* experiences, like emotions or passions, is then required.

3.3 Specificity of the Psychophysical Correlations

A complete elucidation of the nature of psychophysical correlations must also account for their *specificity*. It must first account for the *universal* psychophysical correlations that characterize our species, such as the fact that fear is associated with an increase and not a decrease in blood pressure, and that, in the long term, experiencing negative emotions is associated with a weakening of immunity while experiencing positive emotions is associated with a strengthening of immunity (Rosenkranz et al. 2003). It must also account for the *individual* specificity of psychophysical correlations that characterize a person in connecting their specific psychological and bodily features. The latter finds their expression in the *individual differences* in emotional reactions to the same life situation and in the proneness to contract certain types of diseases -an issue explicitly considered by the psychosomaticians since Alexander's work (Alexander 1950). These individual differences lie in the nature of the experienced emotions but also in their intensity, whose value is clearly context-dependent and can only be assigned within a given range.[12]

[11] The placebo "effect" will be explored in more detail in Sect. 5.3 where it is modelled within the framework of a generalized version of quantum theory presented in Chap. 4.

[12] The indeterminacy in the intensity of emotions, closely related to their complementarity, had been alluded to in Sect. 2.2 and will be explained in detail in Sect. 4.2.

In agreement with Darwin's evolutionary theory of emotions, it would seem that the origin of the species-specificity of these correlations has to be found in the evolution process broadly construed. The establishment of universal psychophysical correlations is essential to prepare the body to deal with primordial emotions such as hunger, thirst, fear or sleep (Lestienne 2008). Primordial emotional experiences have been correlated with biological programs capable of reacting or adapting to them, as illustrated by the hormonal and nervous changes required for preparing the individual to respond to a fearful situation. This first, general category of correlations has indeed been studied for a long time in different philosophical and medical traditions. Traditional Chinese Medicine is based on correspondences between organs and emotions. For example, sadness is associated to lungs, anger is associated to liver and fear to kidneys. These statements, whose probable empirical origins are difficult to find out, do not presently find clear experimental or theoretical assessments (Janssen 2006 pp. 231–233). However, it has been shown on an experimental basis that topographically distinct and culturally universal bodily sensations can be associated with distinct emotional states (Nummenmaa et al. 2013). This has been shown by asking a multi-cultural population to report the intensity of their sensations in different regions of their body when they are subject to emotional stimuli by coloring the regions of body silhouettes. The results of this experiment are reported hereafter (Fig. 3.1):

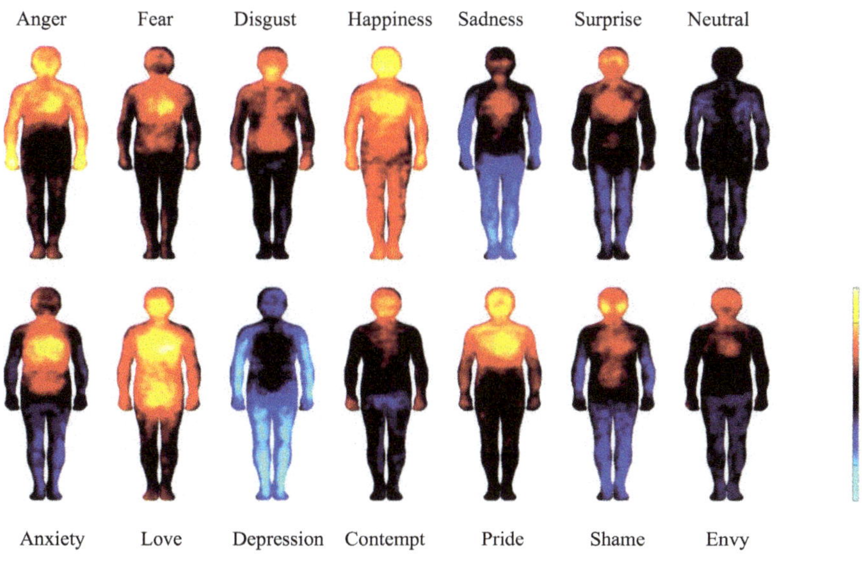

Fig. 3.1 Bodily topography of basic (Upper) and nonbasic (Lower) emotions associated with words. The body maps show regions whose activation increased (warm colors) or decreased (cool colors) when feeling each emotion ($P < 0.05$ FDR corrected; $t > 1.94$). The color bar indicates the t-statistic range. From "Bodily maps of emotions", Nummenmaa et al. PNAS 2013

These universal psychophysical correlations must indeed be regarded as being *dynamical* since they take place between two changes: it is the *triggering* of emotional experiences and the *activation* of specific parts of the body that are strongly correlated. Along this line of thought, Traditional Chinese Medicine teaches us that stimulation by needles of particular areas of the body situated on specific meridians can trigger emotional reactions (Lytle 1993), as illustrated by the treatment of anxiety and depression by acupuncture (Fleckenstein et al. 2018; Luo et al. 1985). These (universal) dynamical correlations between physical actions on the body and emotional changes can also be observed in Shiatsu, a manual therapy technique of Japanese origin, since a pressure on a specific area of the body can trigger emotional discharge (Anderson 2008). They are also very well illustrated in the previously mentioned oriental ancestral practices, like Tai Chi or Qi Gong. For example, statistical studies have shown that the practice of Qi Gong significantly soothes negative feelings, pain in particular, as well as it reduces the level of cortisol in blood and improves immune response (Ryu et al. 1995; Wu et al. 1999).

Let us now turn to the *individual* specificity of psychophysical correlations. This specificity has been explored by psychosomaticians by studying the link between emotional disorders and somatic diseases. Referring to clinical experience, Alexander worked out a theory of the specificity of conflict according to which each emotional disorder would be associated to a specific somatic syndrome disorder (Alexander 1950). He has for example suggested that the psychological conflict that may exist between desires for dependence and independence is associated in the somatic domain with the proneness to develop a peptic ulcer. This association has been tested experimentally on a group of US Army recruits after intensive training. This study has shown that the subjects who had a strong desire for affective dependence while experiencing authority in a confrontational manner are much more likely to develop a peptic ulcer (Weiner et al. 1957) -a result that has been confirmed by a large experimental study conducted in Mannheim (Platz et al. 1993). More generally, systematic studies have been carried out by the School of Chicago in order to highlight the individual specificity of the correlations between experienced emotions and respiratory, cardiovascular, gastrointestinal or dermatological disorders (Alexander et al. 1968). William Reich's "vegetotherapy" (Reich 1933) is based on the existence of direct correspondences between the specific psychological experience of a subject and their anatomy. Reich has defined five personality types endowed with specific anatomic, emotional and intellectual features in connection with five great existential fears. The schizoid type, for example, with long and thin legs, angular anatomy and empty look, arises from the fear of being rejected or from being unable to find a place in an environment deemed hostile. The underlying idea is that childhood emotional traumas engender body tensions which, by repeating themselves, eventually deform the body. Reich and Lowen developed a therapy that focuses on the subject's energy and emotional restructuring, with a view to relaxing their mental and bodily armors and restoring the energy flow, in order to enrich their personality. Along this line of thought, let us once more mention Janov's primal therapy (Janov 2000) that appeals to a concept of "body memory" to emphasize the embodiment of a subject's emotional traumatisms into their body. This embodiment expresses the existence of close correspondences between these specific emotional contents and specific somatic features. Somato-psychic approaches, such as Janov's primal therapy or Fiammetti's somato-emotional therapy previously mentioned, are based on the principle according to which specific parts of the patient's body encapsulate past emotional conflicts that have not been expressed and that an action on these regions can release these emotional conflicts.

The proneness to develop specific somatic diseases has also been successfully associated with the subjects' behaviours, classified in types. Subjects with type A behaviour are characterized in particular by their excessive implication in work and the need to control their environment, their strong desire of perfection, their ambition and their possible arrogance (Friedman and Rosenman 1974). As confirmed by large experimental studies, type A behaviour is significantly correlated to the proneness to develop cardiovascular diseases, like coronary heart diseases with a risk of myocardial infarction (Haynes et al. 1978). More precisely, it has been shown that the so-called "antagonist hostility", a feature of type A behaviour that denotes the difficulty to experience pleasure and by an arrogant attitude, is more specifically associated to cardio-vascular disorders (Littman 1993). In contrast, it has been speculated that subjects of type C behaviour, who are quiet, patient, rather passive and even defeatist, repressing aggressiveness and anger, would have a greater risk to develop cancer (Temoshok and Hefler 1983), while type B behaviour denotes a lower risk of developing a disease.

This behaviourist characterization of somatic disorders does not *explicitly* deal with the question of specificity of *psychophysical* correlations we focus on here. However, these behaviours are accompanied with emotions, feelings, intentions and desires that are actually experienced *subjectively.* Consequently, the observed correlations between behaviour types and proneness to develop somatic disorders could be understood as psychophysical correlations, *between subjective experiences and proneness to develop somatic disorders*. This means that the previously mentioned correspondences between behaviour types and somatic disorders could be very informative for dealing with the question of the specificity of psychophysical correlations provided that reliable descriptions of the purely subjective factors on which the behavioural types rely can be given. Fortunately, this rather difficult enterprise can be avoided. As mentioned above, Broom's "mind-body" practice is grounded on direct correspondences between specific subjective *meanings* built by a patient throughout their life and specific somatic disorders. Subjective meanings are not here confused with their behavioural *manifestation* or even with their linguistic *expression*[13] but are associated as such to specific somatic disorders. These specific correspondences between meanings and somatic disorders for a patient

[13] The study of the way persons talk about their illnesses is the object of "narrative medicine". The latter focusses on the linguistic expression of the complex meanings developed by a patient in regard to their disease, while Broom's "mindbody" medicine deals with the whole symbolic dimension of somatic disorders, which includes their linguistic expression but is more general. Both approaches are of course closely related.

define Broom's notion of *symbolic diseases.* In Broom's words, symbolic diseases occur when:

> the organ system involved, and/or the pathological process, and/or the clinical phenomenology, appears to be particularly congruent with, or appropriate to, the patient's subjective meanings or "story", as ascertained from the patient's language, life history, and behaviours (Broom 2007).

Broom has reported precise cases of patients whose somatic diseases "appear to be symbolic somatic representations" of their personal meanings or individual 'stories' and the way he cared for them. For example, he tells of the case of a woman who:

> developed very aggressive rheumatoid arthritis whilst in an enmeshed marital relationship and social context about which she used the following language: 'in a bind' 'I can't get moving,' 'captured,' 'tethered, enslaved,' and, indeed, 'crippled' by being in the town in which she lived. Therapy addressing these issues has led to remission of 12 years. (Broom 2007).

Broom's semantic approach to the mind-body connection extends Chiozza's linguistic metaphor (Chiozza 1976), which can in turn be traced back to Groddeck's work (1928). It can also be regarded as a refinement of Freud's approach to psychosomatic disease previously mentioned, regarding in particular the "choice of organ" taking place during the "somatization" or "conversion" process of an unconscious psychic conflict (Freud 1905/1990). Along this line of thought, the specificity of psychophysical correlations is thus of *semantic order*: it expresses the specificity of the correspondence between a concrete signifier, which is here the somatic disorder under consideration, and the specific signified the person attributes to this signifier. This signified is the meaning that the subject associates to their illness in regard to their personal experience of life, which involves a specific socio-cultural context and, in particular, the social relationships established with others. This semantic approach to the specificity of psychophysical correlations will be further developed in Chap. 5 with regard to healing.

Finally note that in addition to these clinical approaches developed in the field of psychosomatic, a finer insight of the individual specificity of the psychophysical correlations can be provided by referring to *embryogenesis*. Embryogenesis describes the organism's development process from the cell to the formation of an autonomous living being. In this process the embryonic leaves (endoderm, mesoderm, and ectoderm) at the origin of the embryo's tissues develop *conjointly* to its first experience of primordial emotions, such as those related to the impossibility of breathing, eating or swallowing, which consequently "shape" their development. The study of embryogenesis could thus inform us about a person's specific psychophysical correlations relating the emotions they feel and the functioning of their organs. For example, as shown by cancerologist Hamer (2005), the cells of the organs responsible for digestion or breathing, which are essential for the survival of the individual, originate from the endodermal leaflet and depend on the brain stem - the oldest part of the brain. Consequently, there would be specific correlations between a subject's primitive emotions associated with breathing, eating or swallowing and the development of the organs governed by the brainstem and derived from the embryonic endoderm such as the lungs for breathing. The search for the origin of the specificity of psychophysical correlations in embryogenesis, which has been extensively developed by Hamer (2005), can provide an essential information on the genesis and development of internal diseases. This point will be developed in Chap. 5 that deals with healing.

3.4 Mind-Body Entanglement

3.4.1 Which Systemic Approach for Dealing with the Mind-Body Interdependence?

The strong interdependence of the mental and somatic features of a person that can be observed in emotions and in any other psychophysical experience suggests the development of a systemic approach to the individual. Is the systemic approach developed by Bertalanffy (1968) and enriched by cybernetic considerations (Wiener 1962) appropriate for modelling the mind-body interdependence?

Bertalanffy's systemic approach is based on the idea that common principles and structures could be found for all types of systems, being physical, biological, sociological or political, by exploiting formal similarities between various scientific models. It explores a holistic approach to phenomena that emphasizes the relationship between the sub-systems and uses the concepts of network, multiple interactions, circular causation, positive or negative retroaction, stability and instability. This holistic view of the world thus contrasts, or at least complements, the mainstream analytical and reductive approach of science that tries to explain the world by mechanisms decomposable into its "last" elements. Bertalanffy's systemic approach and its cybernetic refinements has applications in various fields. For example, in the field of biology, modelling the maintenance of life according to this approach leads to the description of the individual as an open system whose complex causal interactions between its parts are governed by the balance morphogenesis/homeostasis (Bertalanffy 1968, Chap. 7). In the field of the human sciences, particularly in psychiatry (Bateson 1979) and psychology (Watzlavick et al. 1967), this systemic approach has proved very fruitful for the development of family therapy. The Palo Alto Mental Research Institute developed with Don Jackson and Paul Watzlawick a comprehensive approach to the problem experienced by the patient, by considering that the symptoms result from a dysfunction of the entire social group they belong (Jackson 1968).

However, if this "classical", systemic approach is very fruitful for dealing with systems whose relationships between parts can be interpreted in terms of *causal interactions*, even in a sophisticated way (circular causation, positive and negative retroaction…), it cannot be applied to the psychosomatic unity. First, the concept of causation cannot be used to describe the relationship between thoughts, emotions or memories. As is well known by psychologists and psychoanalysts,[14] *their dynamics*

[14] See, for example, Freud's *Metapsychology* (Freud 1946).

are governed by laws of associations, projections, shifts and metaphors and not by laws of cause and effect that have been successfully applied in the physical and biological domains.[15] A paradigmatic illustration of the dynamic of emotions is provided by the sudden emergence of an emotion without apparent causal connection with the flow of thoughts, neither with the biological and social context. This emergence is based on the *association* in the subject's memory of the different components of a past emotional experience (regarding, for example, its circumstances or its actors) which is then reactivated when any of these associated components is encountered again afterwards (Leventhal and Scherer 1987; Lang 1979). Second, it seems that this "classical" approach cannot grasp the very foundation of the psychosomatic unity of the individual. As explained above from the study of emotions and already from the logico-philosophical analyses of Chap. 1, no clear temporal order or causal connection can be consistently defined between subjective experiences and their somatic correlates. Such a unity seems to be grounded on the existence of specific *a-causal* psychophysical correlations.[16] Moreover, as mentioned in the discussion of Spinoza's psychophysical parallelism (Sect. 2.2), psychophysical correlations are non-local: they cannot be conceived as totally predetermined despite the evolutionary constraints, due to the indeterminacy of mental properties (particularly the intensity of emotions),[17] allowing thus imagination and free will. This means that a new systemic approach capable of accounting for the existence of a-causal and non-local correlations is required for dealing with the mind-body connection. We will see that such a systemic approach must be grounded on the concept of generalized entanglement.

3.4.2 *Mind-Body Entanglement, Condition of Possibility of the Psychosomatic Unity*

The concept of *mind-body entanglement* grasps the previously highlighted properties of the correlations between a person's mental and bodily features. This concept will find a precise, mathematical representation within a quantum-inspired framework (see Chap. 4). However, we will provide here an intuitive definition of this concept, illustrated by Fig. 3.2 below, and explain why mind-body entanglement is the condition of possibility of the psychosomatic unity. Some of its consequences, further explored in more detail in Chap. 5, are also mentioned.

---

[15] The a-causal dynamic of psychism has already been mentioned in Sect. 2.2 with regard to questioning the generally accepted interpretation of Spinoza's mind-matter parallelism in terms of a strict isomorphism.

[16] The reader is reminded that Wegner's experiments and their extensions mentioned in Chap. 1 show that the study of the nature of the mind-body connection can be skewed by our (possibly misleading) feeling of a causal responsibility of the mental. In view of this result, of the logico-philosophical analysis of Chap. 1 showing the inappropriateness of the concept of causation to explain the mind-body connection and of the previous considerations on emotions, it may be easier for the reader to accept this conclusion.

[17] This indeterminism will also be shown for physiological properties in Sect. 4.

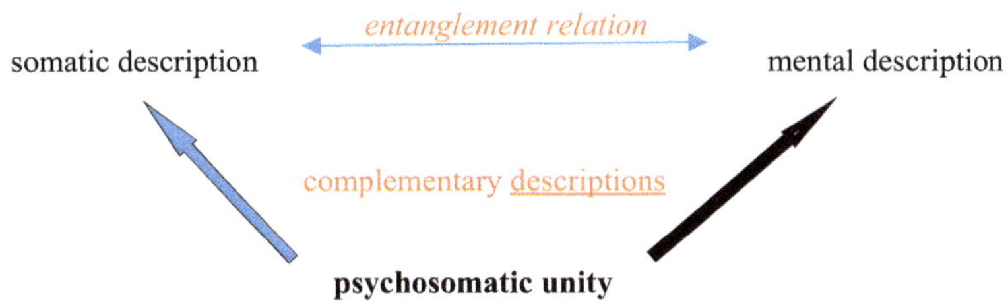

Fig. 3.2 A schematic illustration of the concept of mind-body entanglement

Properties of the mind-body connection:

- *Inseparability of the somatic and the mental aspects of the psychosomatic state* The concept of mind-body entanglement first expresses the idea according to which the mental and the somatic states of the individual are strongly *interdependent*, that one of them cannot be described independently of the other. This strong mind-body interdependence reflects the fact that the individual is *an indivisible psychophysical unity* whose description cannot be reduced to the simple adjunction of a somatic description and a mental description since the information relative to the somato-psychic correlations constitutes an *essential and irreducible element of this global description.* This *inseparability* of mental and somatic features of the individual, which indeed closely relies on the non-locality of their correlations (see below), has been explored above (Sect. 3.1) from Descartes' concept of union of mind and body and from the developments of psychosomatic, and it has been illustrated in the case of emotions. *It will be expressed in Chap. 4 by the non-factorability of the state vector representative of the psychosomatic state of the individual.* As a consequence, any action performed in one of the mental or somatic domain can affect the other *without any signal transfer and regardless of any spatial or temporal localisation.* This non-signaling property of the mind-body connection is one of the fundamental principles underlying the holistic approach to healing presented in Chap. 5.

- *Specificity (or meaningfulness) of psychophysical correlations* The inseparability of mental and somatic features is of course *specifically defined for each person*. Our analysis of the question of specificity of psychophysical correlations in Sect. 3.3 has led us to assert, in agreement with Broom's work, that the genuine nature of psychophysical correlations is of *semantic order*. Similar to the relation between the *signifier, which is the* material form of a sign that can be heard, and the signified, which is the mental concept associated to this signifier by a speaker, psychophysical correlations are correlations between *a subject's somatic features and elements of signification*[18] *that they assign to a life*

[18] Of course, in agreement with the multiple realizability argument against type-identity theory (see Chap. 1), these elements of signification cannot be considered as being rigidly associated with bodily features.

situation. The psychophysical states of the individual thus put in correspondence bodily features and meanings *for a subject and in a particular context*. Their relationship, analog to that between signifier and signified in linguistic, is of *semantic order* and has a priori nothing to do with a "cause and effect" connection. As explained in Sect. 3.3 and in more detail in Chap. 5, the establishment of such meaningful correlations, encoded in the entangled psychophysical state of the individual, results from a long "preparation" process (evolution process and social life). Taking into consideration this preparation process explains the existence of a body memory and allows a deeper understanding of the origin and evolution of internal diseases (see Sect. 5.2).

These two first properties of the mind-body connection have been observed for a long time, from Greek physicians to the advent of psychosomatic. However, the next two are quantum-like properties of psychophysical correlations whose fundamental importance and consequences has not yet been fully understood. These two properties characterize the relation of generalized entanglement that will be rigorously defined in the next chapter:

- *A-causality of psychophysical correlations.*
  The a-causality of psychophysical correlations means that no causal relation, whatever its nature, can consistently explain the strong interdependence between the somatic and the mental features of a person. More precisely, these correlations can neither be understood as an exchange of signals or a transfer of information (they are 'non-signaling'), in accordance with the processual approach to causation mentioned in Chap. 1, nor, more generally, as "a generation, or effective production and determination" (Kim 2005). The alleged existence of a mind-body causal interaction has first been questioned from Libet's work and its recent extensions and by the logico-philosophical analysis of Chap. 1; it has also been questioned on experimental grounds from Canon's theory of emotions (Sect. 3.1). It has been concluded that the concept of mental causation suffers from at least one of the following issues: lack of specificity (when it reduces to that of correlation, like for example in the counterfactual approach to causation); logical inconsistency (within the framework of the non-reductionist physicalism) or, within dualism of substances, by its lack of experimental evidence and its probable incompatibility with current physics. Moreover, as will be explained in Sect. 4.4, the concept of psychophysical causation can be refuted experimentally.

- *Non-locality of psychophysical correlations*
  The non-locality of psychophysical correlations means that these correlations cannot be explained by the total predetermination of the correlated entities. As explained above (Sects. 2.2 and 3.2) and in more detail in Chap. 4 below, measuring mental or somatic observables does not provide values intrinsically possessed by these observables before the measurement because measuring them change the subject's psychosomatic state. This fundamental indeterminacy in the values of mental and somatic observables, which is closely related to their

complementarity (intuitively, their partial incompatibility[19]), gives rise to the property of non-locality of psychophysical correlations. Non-locality of correlations is a typically non-classical property since it means that no-signaling correlations cannot be explained, like in classical physics and according to common sense, by the intrinsic properties of the correlated entities. It is here the direct expression of the irreducible indeterminacy in the corresponding mental and somatic observables that can never be completely eliminated.

The concept of mind-body entanglement thus makes it possible to go beyond the concept of individual as a mere "union" of two independent parts, mental and somatic, whose properties either "interact" according to certain unknown laws or would be intrinsically predetermined before any observation made on him. Moreover, it allows the representation of the specificity of psychophysical correlations, which relies on semantic considerations about the "preparation" of the individual and their social life and not on impersonal (or even "mechanical") "cause and effect" laws. We have seen that describing emotions and all global, psychophysical phenomena experienced by a person, like drives or feelings, requires all these conditions to be satisfied. Consequently, *mind-body entanglement is the condition of possibility of the psychosomatic unity of the individual.* What makes this unity possible is the relation of generalized entanglement that links its psychic and somatic aspects and whose properties meet the four requirements mentioned above, namely, inseparability, specificity, a-causality and non-locality. The entanglement of a person's psychosomatic states can be conceived of as the expression of their organizational closure since the latter expresses the way mental and somatic features of the individual are correlated. For example, the way their emotions are correlated with the endocrine processes that govern the functioning of their organs takes an important part in their psychosomatic organization. The concept of mind-body entanglement can explain some paradigmatic psychosomatic phenomena whose classical interpretation, in terms of causation, remains totally mysterious. It is the case of the placebo "effect", biofeedback and mental imagery phenomena, as well as the "effect" of meditation on the body. These phenomena will be presented and modelled in Chap. 5 along our systemic, quantum-like approach.

Another striking consequence of the mind-body entanglement is that, by the very nature of the entanglement relation, the specific psychophysical correlations are time and space independent. The space independence of the mind-body connection could explain the holographic models of the body developed by traditional medicines (see Sect. 5.2). In addition, this property is the fundamental ingredient of an explanation of distant healing phenomena explored in Sect. 5.4, where the healer and the sick are not next to each other. As to the time independence of the mind-body connection, it means that the somatic features of a subject can be strongly linked to their emotions *regardless of any temporal separation.* This property explains why a bodily dysfunction can be connected to past or future thoughts and

[19] The complementarity of mental observables and that of somatic observables will be explored in detail within an appropriate mathematical framework in Chap. 4.

feelings. In particular, it can explain the presence in a person's psychosomatic state of the emotional conflicts endured in their past. This phenomenon is generally interpreted by psychoanalysis as the signature of the Freudian "unconscious" but it finds here a different interpretation, in terms of time entanglement. This interpretation grounds the very possibility of restoring the patient's body memory by acting on their body, as was illustrated in Sect. 3.1 by Reich's vegetotherapy, Janov's primal therapy and Fiammetti's somato-emotional approach. This essential point will be further developed in Sect. 5.2. Conversely, the time independence of the mind-body connection provides a possible justification of Libet's and Wigner's experimental results previously mentioned (Chap. 1) according to which a significant time interval can separate the activation of the brain ("readiness potential") from the conscious decision (Chap. 1). It also leads to a consistent explanation of the strange presence of past bodily injuries in the present moment, observed for example in the "phantom limb" pain which occurs when a limb that has been amputated still feels like it is attached to the body and gives rise to strong pain.

As explained in Introduction, the concept of mind-body entanglement, illustrated by Fig. 3.2, is built in two steps. First, by generalizing the concept of entanglement of quantum physics in order to make it consistent and testable for general systems, and, second, by applying it to the mind-body connection. This construction will be made more precise within the mathematical framework of a generalized version of quantum theory, in the next Chap. 4).

Chapter 4
A Quantum-like Representation of the Psychosomatic Unity

In this chapter, the psychosomatic unity of the individual and its condition of possibility, mind-body entanglement, will be represented within a theoretical framework that provides the required conceptual and mathematical tools. As shown by Bohm and Hiley (1993), by Atmanspacher et al. (2002), by Primas (2003, 2010) and, more recently, by the author of this book (Uzan 2014a, b, 2016, 2017, 2020), the key quantum concepts of complementarity and entanglement can be exported beyond the strict framework of quantum physics and used to model many phenomena, in particular all mind-related phenomena which are not governed by local causation and classical logic. For example, the way we successively respond to two questions, we combine concepts in a sentence or we make a decision in uncertain situation have been successfully modeled within this quantum-like framework. The reason of this success is that the latter theoretical framework allows taking into account the order and interference effects that are inherent to these cognitive acts (Busemeyer and Bruza 2012; Aerts et al. 2011). Following this line of thought, this chapter presents an approach to the notion of psychosomatic unity within a generalized version of quantum theory that keeps its algebraic structure but where all a priori references to the physical world have been relaxed. In such a generalized systems theory, the notion of "system" is defined as any part of reality which is an object of attention and investigation and that of "observable" as any meaningful property which can be measured or observed in a reproducible way. Within this theoretical framework the notions of generalized complementarity and generalized entanglement can be conveniently represented, and it becomes possible to conceive an *experimental test* of the nature of the mind-body connection.

The original version of this chapter was revised. The correction to this chapter is available at https://doi.org/10.1007/978-3-030-90173-8_7

P. Uzan, *Mind-Body Entanglement*, SpringerBriefs in Philosophy,
https://doi.org/10.1007/978-3-030-90173-8_4

4.1 Quantum-Like Representation of the Psychosomatic States

The theoretical framework where the psychosomatic states of the individual can be conveniently represented is a generalized form of quantum theory. Such a generalized version of quantum theory keeps its fool mathematical formalism but deletes any a priori reference to the physical world. The reason of this choice is that, as will be shown in detail hereafter, the underlying algebraic formalism of quantum theory is a non-commutative C*-algebra of operators (and not of functions, like in classical physics) capable of representing elegantly and fruitfully the complementarity of observables and the entanglement of the state of a composite system. Appendix 1 provides an intuitive presentation of quantum theory, namely a simplified definition of C*-algebra and of the corresponding Hilbert space formalism to deal with all possible states of a system; it also explains that the state-space of a composite system must be defined as the tensor product of the Hilbert spaces of its sub-systems.

In the present situation, the generalized system is the individual whose psychosomatic state is represented by a vector in a Hilbert space, which is a complex vector state endowed with a scalar product. Similar to the state space associated with a physical system composed of two parts in quantum theory, the space of the psychosomatic states of the individual is defined as the tensor product of a Hilbert space M of mental vector-states and a Hilbert space S of somatic vector-states:

$$H = M \otimes S.$$

This notation means that a psychosomatic state of the individual is conceived as a linear combination of possible "product-states" associated with two values of elementary observables respectively defined in the somatic and the mental domain. By "elementary observables" we mean the basic quantities that are usually considered and measured in both mental and somatic domains. For the somatic domain these quantities are physico-chemical parameters like the temperature of the body, the blood pressure or the blood concentration of different cells, platelets and other substances (like hormones or nutrients). More complex measurable quantities, like the velocity of metabolism or the degree to which an organ dysfunctions, can be defined from the latter ones that are taken as primitive. For the mental domain, the primitive quantities are those identified by psychology and measured by psychological tests, like intensity of felt emotions, emotional sensibility or even personality, from which can be defined more complex quantities, such as the degree of depression or social phobia. It will be assumed that a finite number of such mental and somatic primitive quantities is sufficient to define the psychosomatic state on an individual[1] and that their possible values can a priori vary continuously. The

[1]This hypothesis about the finite number of elementary patterns from which the state of the individual can be defined has always been adopted and has proved to be fertile, just like the four humours theory of ancient Greece, the finite number of morphological types and associated tem-

quantities that can be measured are represented, like in quantum theory, by self-adjoint operators[2] acting on H and whose eigenvalues, which are real numbers, are their possible values. The spaces S and M are thus spanned, respectively, by the eigenstates of the somatic observables and by the eigenstates of the mental observables that are used for describing a person's psychosomatic state. These eigenstates respectively define a "natural" or "elementary" basis for the somatic and the mental domain.

Note that the representation of the state space H by a tensor product of a somatic state space and a mental state space does not a priori assert anything on the nature of the psychophysical correlations: they can still be causal interactions, they can be common cause, classical connections or a-causal and non-local entanglement correlations. In the latter case that will be made more precise in Sect. 4.3, the psychosomatic state of the individual is *non-factorable,* that is, it cannot be written as the tensor product of a mental state and a somatic state.

Formally, a (non-necessary orthogonal) basis of unit vectors of H is the tensor product of the elementary M-basis and S-basis and can then be written, in Dirac's notation, as:

$$\Pi = \left\{ |m_k, s_{k'} > \right\}$$

where the indexes k and k' refer, respectively, to these somatic and mental primitive quantities (which are in finite number). A vector $|m_k, s_{k'}> = |m_k> \otimes |s_{k'}>$ of Π is the tensor product of the eigenvector $|m_k>$ associated with the (supposed non-degenerated) real eigenvalue m_k of the mental observable $\mathbf{G}_M$ of M, and of the eigenvector $|s_{k'}>$ associated with the (non-degenerated) real eigenvalue $s_{k'}$ of the observable $\mathbf{J}_S$ of S.

Along this quantum-like approach the (normalized) psychosomatic state of an individual will then be decomposed on the basis Π as:

$$|\Xi>= \iint_{mk,sk'} c(m_k, s_{k'}) |m_k, s_{k'} > dm_k \, ds_{k'},$$

where the complex coefficients $c(m_k, s_{k'}) = < m_k, s_{k'} | \Xi>$ are defined as continuous and derivable functions of the continuous variables m_k, and $s_{k'}$. Moreover, one will introduce the only possible probability measure on Hilbert subspaces (Gleason 1957), asserting that:

(Pr) If A and B are, respectively, a somatic and a mental observable, then $|< a,b|\Xi> |^2$ is the probability to measure A = a and B = b on the system in state $|\Xi>$.

---

peraments proposed by Sheldon (1942) or the finite number of structural "items" and mental "data" in the psychoanalytic classification proposed by Marty (1963).

[2] The representation of mental and somatic observables by operators is justified by the fundamental indeterminacy of their values whose consequence is that only a range of possible values can be assigned to these observables before measuring them (see Sect. 3.4). This essential point will be clarified when the notion of complementary mental and somatic observables is explored (Sect. 4.3).

The adoption of this rule, which implies the normalisation condition:

$$\iint_{mk,sk'} |\langle m_k, s_{k'} | \Xi \rangle|^2 \, dm_k \, ds_{k'} = 1,$$

means that (in the Π-representation) the psychosomatic states form a sub-ensemble of L^2, the space of square summable functions.[3] We thus work in a mathematical structure quite similar to that of quantum theory. Also note that the transition probability amplitudes $< m_{k1}, s_{k'1} \mid m_{k2}, s_{k'2} >$ between the elementary product-states $|m_{k1}, s_{k'1} >$ and $| m_{k2}, s_{k'2} >$ are a priori different from zero since these product-states are not necessarily orthogonal. This expresses the fact that emotions we experience are generally not exclusive (for example, we can experience sadness and anger at the same time) and that, similarly, different hormonal activities co-exist.

For sake of simplicity, we will suppose in the following of this book that the set of possible values of all the somatic and mental observables form *discrete ensembles* and, consequently, define discrete ensembles of corresponding eigenstates. This simplification is justified by the fact that the values we can measure are always subject to some minimal experimental uncertainty, whatever the precision of measurement. Consequently, their continuous spectrum of values, if any, can be discretized into a finite number of intervals whose size is determined by this minimal uncertainty. With this simplification, a person's psychosomatic state can be written as the following *discrete sum* in the basis $\{| m_a, s_b>\}$ as:

$$|\Xi >= \Sigma_{a,b} \, c(m_k, s_{k'}) | m_k, s_{k'} >.$$

Let us also introduce a simplification of writing, often very useful. According to the particular observables we focus on, the terms of the psychosomatic state can be grouped so as to emphasize the feature(s) we want to focus on. For example, if we want to focus on the nature of a dominant emotion, called "J", we can define unit psychosomatic states $|J+,s+>$ and $|J-,s->$ of the space ***H*** as, respectively, the sum of the terms of the psychosomatic state $| \Xi >$ for which the intensity of emotion J is significant (say, bigger than a certain value) and not significant (smaller than this value). To illustrate this very useful way of writing, which will be used in Chap. 5, suppose that we want to focus on the emotion "sadness" and that we consider that this emotion is dominant if its intensity j on a scale ranging from 0 to 10 is bigger than 7. A person's normalised psychosomatic state will have the following form:

$$|\Xi >= \alpha_+ |J+, s+> + \alpha_- |J-, s->,$$

[3]The latter assertion is essential since, as will be shown in the following of this chapter, it will allow us to say that any psychosomatic state in the Π–representation can be decomposed as a weighted sum of harmonic functions, which makes possible the mathematical definition of couples of complementary observables in the somatic domain.

where $|\alpha+|^2 + |\alpha-|^2 = 1$ and the partial sums $|J+, s+>$ and $|J-, s->$ are defined as:

$$|J+,s+>=(1/N+)\ \Sigma_{j,k}\ c(j,k)|m_j : j>7, s_k>.$$

$$|J-,s->=(1/N-)\ \Sigma_{j,k}\ c(j,k)|m_j : j\leq 7, s_k>,$$

N+ and N− being, respectively, the norm of these two vectors.

4.1.1 *Temporal Evolution of Psychosomatic States*

As shown by chronobiology and chronopsychology studies, most of the physiological and psychological features of the human are governed by internal rhythms which are synchronized with each other and with the day / night cycle imposed by the sun (Reinberg and Smolensky 1983). All bodily and mental observables are changing periodically, according to periods which can be classified as circadian (of approximately 24 h, synchronised on the day/night rhythm), ultradian (less than 24 h) and infradian (longer than 24 h), the regulating clock of these rhythms being located in the hypothalamus. In the somatic domain, one can mention for example the circadian rhythm of heart rate, body temperature or secretion of hormones by the endocrine glands, such as the secretion of melatonin which regulates the cycle sleep/wake, or the infradian period of menstruation. In the mental domain one can for example mention the circadian rhythm of learning or that of the emotional sensibility (Beugnet-Lambert 1988). Each important biological function and each organ has indeed its own rhythm, intended to optimize the efficiency of the involved local structure according to the environmental context (INSERM 2018). However, in ideal case, these local biological rhythms are synchronized with each other and with the circadian rhythm in order to ensure a good functioning of biological and mental functions. For example, the sleep/wake cycle is synchronized with the night/day cycle and governed by the synchronized secretion rhythms of melatonin, which increases until a peak around 3 am and then gradually decreases until awakening, and is also synchronized with the cycle of cortisol (with a minimum in the middle of the night and maximum between 6 a.m. and 8 a.m) and with that of the ACTH hormone, whose concentration is maximal just before the peak of cortisol.

In the ideal case of perfectly synchronized biological and psychological rhythms, the components c_j of the psychosomatic state, which are complex probability amplitudes (see above), change harmonically as:

$$c_j(t)=c_j(0)\exp(-i\,\omega_j t),$$

with $\omega_j = 2\pi/T_j$, where T_j is the period common to the emotional and somatic observables that define the product-state considered. This implies that the rhythmic evolution of psychosomatic states is governed by a unitary operator:

$$\mathbf{R}(t+\Delta t, t) = \exp[-i\Omega\Delta t],$$

where Ω is an hermitian operator whose associated matrix (always in the Π-representation defined above) is diagonal, its diagonal terms being equal to ω_1, $\omega_2, \ldots, \omega_n$, which means that the basic vectors of H are the eigenvectors of Ω with (non-degenerated) eigenvalues ω_j: $\Omega\, |\Xi_j> = \omega_j\, |\Xi_j>$.

In the case all these internal rhythms are not synchronized with each other, and thus not synchronized with the "conductor", namely the circadian rhythm, the probabilities of occurrence of some of the elementary product-states can be very small. This is because combining two dephased waves of same frequency gives rise to interference phenomena, which can be destructive,[4] or/and beats in the case of slightly different frequencies. As will be clarified in Chap. 5, the psychosomatic imbalance that originates illness can be explained by this lack of synchronisation.

4.2 Complementarity in Quantum Physics and Beyond

In quantum physics, the notion of complementarity has been implemented according to two different ways. In a first sense, it applies to the *descriptions* of a same quantum system that exclude each other but are nevertheless required together to give a complete information on it. An enlightening illustration of the notion of complementarity of descriptions is provided by the paradigmatic wave-corpuscle complementary aspects of light in Young's two slits experiment: light can be modelled as being composed of corpuscles (photons) by observing their impact on a screen, but light can also be described as a wave since it gives rise to wave phenomena, as is the case for the interference pattern that can be observed on the screen.

The notion of complementarity of descriptions can be rigorously characterized in algebraic terms. A "description" can be mathematically defined as an algebra of experimental propositions susceptible to describe the phenomenon under study and endowed with logical connectors (and, or, not,..) governed by logical rules. A description governed by the laws of *classical* logic (like the law of excluded middle or the distributivity of the "and" over the "or") defines what is called a Boolean algebra. The notion of complementarity of descriptions can thus be defined as follows (Primas 2010):

> *Two descriptions are "complementary" if they cannot be embedded into a single Boolean algebra.*

In other words, two complementary descriptions cannot be combined in order to form a single description governed by classical logic. This definition can be illustrated by the wave-corpuscle complementarity of light mentioned above. In Young's

[4] As will be explained and illustrated in Chap. 5, the de-synchronization of internal rhythms each other and with the circadian rhythm grounds the development of illness within our systemic approach.

double slit experiment, the complementary descriptions of light are, respectively, relative to different experimental contexts. When both slits A and B are open (Fig. 4.1), one can observe an interference pattern on the screen, the position of the particles being in this case *only detected on the screen.* This wave-like description can be done in a first experimental language L_{AB}.

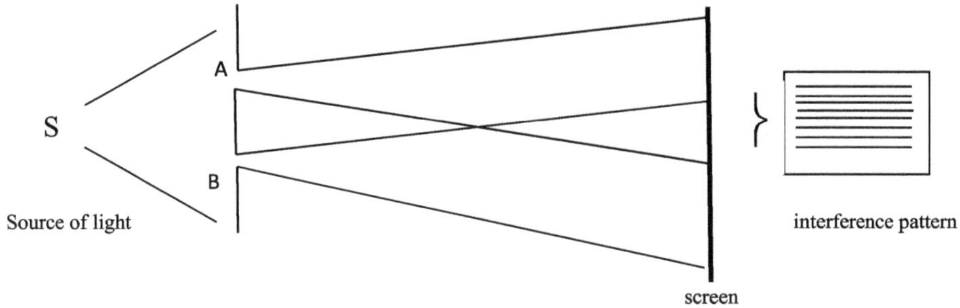

Fig. 4.1 Young's double slits experiment

However, one can determine which slit the particles pass through by placing counters at the slits. If we can know with certainty which slit the photon passed through, the interference pattern disappears, which means that we can only observe the corpuscular aspect of light.[5] In fact, three different similar experiments can be performed depending on which slit is open (only A, only B or A and B). These three experiments can define three corpuscular descriptions L_A, L_B and (L_A v L_B) which respectively describe the passage of the photon through A, through B or through one of the two slits (Bitbol 1996, § 1.2.10). If the corpuscular descriptions L_A, L_B and (L_A v L_B) generate a Boolean algebra, it can be shown that *the wave-like description L_{AB} and the corpuscular descriptions cannot be embedded into a single global Boolean algebra* (Bitbol 1996, annex 1), which expresses the complementarity of corpuscular and wave descriptions.

Let us now focus on the related notion of complementarity *of observables* in quantum physics and on its possible generalisation. In quantum physics, two physical quantities A and B are said to be "complementary" if the corresponding operators **A** and **B** which represent them mathematically do not commute:

$$[\mathbf{A},\mathbf{B}] =_{df} \mathbf{AB} - \mathbf{BA} \neq 0$$

This relation means that the sequence in which A and B are measured is not indifferent. If A and B are complementary observables and that B is measured first, the measurement of A performed after that of B disturbs the state of the system and erases the previous value of B, as can be shown by a second measurement of B. This property can be illustrated using three Stern-Gerlach devices that successively

[5] Note that if we do not know with certainty which slit the photon passed through, the visibility of interference at the screen is only reduced. In this intermediate case, both aspects of light co-exist. For the sake of simplicity, we only focus here on the extreme cases.

measure the spin of a particle according to two different directions (respectively, for S_x and S_y), which defines two complementary observables. It can be shown that the statistical distribution of the results of measurement of S_x and S_y depends on the order they are measured (Cohen-Tannoudji et al. 1973, Chap. IV).

The non-commutativity of two observables can also be seen by the impossibility of measuring them simultaneously with accuracy, a joint measurement showing that their possible values are necessary scattered around their mean value. In this case, it can be shown that their standard deviations, which cannot be interpreted as experimental uncertainty but are inherent to their very definition as operators, are related to their commutator by the following indetermination relation, which holds for any state of the measured system:

$$\Delta A\, \Delta B \geq 1/2 |< [\mathbf{A},\mathbf{B}] >|,$$

where $< [\mathbf{A}, \mathbf{B}] >$ is the expectation value of their commutator in the considered state.

For example, the position observable **X** and the momentum (in the x-direction) observable $\mathbf{P}_x$ do not commute:

$$[\mathbf{X},\mathbf{P}_x] = i\hbar \mathrm{I},$$

their degree of non-commutativity being evaluated by ħ, the Planck constant divided by 2. They satisfy the following indetermination relation:

$$\Delta X\, \Delta P_X \geq 1/2\hbar.$$

Note that the two presented approaches of the notion of complementarity of descriptions and of observables are indeed closely related since the complementarity of descriptions relies on that of the observables that are used for these descriptions. For example, in the previously mentioned case of the wave-corpuscle complementary descriptions of the double-slit experiment, the observable that determines the slit the particle has passed through, which is involved in a corpuscular description, does not commute with the observable that measures the position of the particle on the screen, which describes the interference pattern.

4.2.1 Complementarity of Generalized Observables

Bohr, one of the founder of quantum mechanics, has suggested to use the notion of complementarity in *any* area of reality in order to characterize the relationship between excluding descriptions which have to be considered together in order to obtain a complete information about a phenomenon. For example, he used the notion of complementarity to characterize the relationship between the mechanistic and teleological aspects of life, which respectively involve causality and finality principles. In psychology, he used this notion to characterize the relationship between the mental attitude of spectator and that of actor, which exclude each other but are both necessary in life. As mentioned previously, the notion of complementarity has been generalized within the framework of a generalized version of quantum theory whose basic ideas and formalism was developed by Hartmann Römer and presented in the paper by Atmanspacher et al. (2002). In this generalized version of quantum theory, *all references to the physical world have been relaxed*, which means that it does not a priori refer to physical quantities, like energy or spin, that Planck's constant (which delineates the border between *quantum* and classical physical phenomena) does not play any role and that the temporal evolution of generalized systems is not a priori governed by the Schrödinger equation (which governs that of quantum systems). As entioned above, a "system" is here defined as *any* part of reality which is an object of attention and investigation and an "observable" can denote *any* meaningful property which can be measured or observed in a reproducible way.[6]

This general version of quantum theory can of course be used for dealing with physical systems, for example to establish the complementarity of the Hamiltonian-like description and the informational description (regarded as an extra-physical description) of chaotic systems (Atmanspacher and Scheingraber 1991). The complementarity of these two modes of dynamical descriptions is expressed by the non-commutativity relation:

$$\mathrm{i}\left[\mathbf{L}, \mathbf{M}\right] = \mathbf{K}\,\mathrm{I}.$$

where **L** is the Liouville operator, which governs the Hamiltonian evolution, **M** an operator which computes the amount of information that can be gained by measuring an ensemble of key quantities on this system and **K** is the Kolmogorov-Sinaï entropy (a positive invariant for chaotic systems).[7]

Most importantly for our purpose, this generalized version of quantum theory can be used successfully for dealing with all mind-related phenomena. For example, in the field of psychology of perception Atmanspacher, Filk and Römer (2009) have considered the Necker cube for dealing with the phenomenon of bistable perception. The Necker cube is a two-dimensional representation of a cube which can be seen from above (a), front face up and at left or from below (b), front face down and at right, and can thus be modelled as a generalized quantum system oscillating between two non-stationary states (a) and (b) and submitted to observation (operator O) (Fig. 4.2):

[6] These authors also suggest to weaken the algebraic structure of standard quantum theory. However, such a weakening is not appropriate for our purpose and will not be retained.

[7] Note that contrary to the non-commutativity of observables in quantum mechanics, which always involves the Planck constant, the commutator [L, M] is not universally determined here since K is system- and parameter-dependent

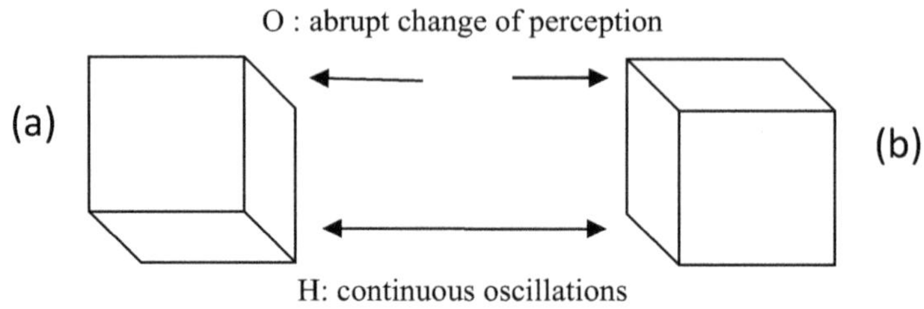

Fig. 4.2 The two possible perceptions (**a**) and (**b**) of a Necker cube

These authors have shown the complementarity of the continuous, non-disturbed, Hamiltonian-like oscillation between the two possible perceptions of the cube (a) and (b), and the abrupt change of perception governed by an observation operator O whose eigenstates are (a) and (b).

In the field of cognition and decision theory, important advances have been realised within this generalized version of quantum theory. Aerts and his team (Aerts et al. 2011) as well as Busemeyer and Bruza (2012) have proposed to model cognitive processes using this generalized version of quantum theory in order to account for the effects of order, contextuality and interference they give rise to. For example, the signification of the conjunction of two words crucially depends on the order they are taken into account, the first one playing the role of a context in which the second one is interpreted. To illustrate this property of cognitive processes, these authors have referred to Hampton's work who showed that, for a random population, the probability P (f and p) that a subject considers garlic as a "food plant" is greater than the maximum of the two values P (f) and P (p) of considering, respectively, garlic as a "food" and "garlic" as a "plant", which violates the classical probability calculus and requires the quantum probability calculus associated with the algebraic structure of generalized quantum theory (Aerts 2009; Aerts et al. 2011). Another important example is provided by the resolution of Ellsberg's paradox relative to the behaviour of economic agents (Ellsberg 1961) within this generalized version of quantum theory. This paradox of the classical decision theory is grounded on Tversky's and Shafir's experimental evidence of a significant violation of the sure thing principle according to which (Savage 1954):

> "If A is preferred to B when X is achieved and if A is preferred to B when not X is achieved, then A is preferred to B even if it is not known whether X is achieved or not."

These authors have shown that Ellsberg's paradox, which questions the validity of the law of total probabilities of the classical calculus of probabilities, can be resolved if we represent the mental states and processes involved in a generalized quantum framework. The reason is that in such a framework it is possible to show that when the decision is taken in an uncertain situation, the difference between the value of the classical probability (to prefer A to B) and the experimental result can be calculated *from the interference term* between a participant's possible mental states of knowledge. More details on this resolution can be found in the references (Aerts et al. 2011; Busemeyer and Bruza 2012; Uzan 2014c).

More generally, Römer (2006) has mentioned the relation of complementarity between substantial and processual observables in *any* domain of reality: *substance* observables refer to timeless features of a system, like the possible states in which it can be found, referring thus to Parmenidean philosophy, whereas *process* observables are relative to its change, to the transition between these states and refer to Heraclitean philosophy. A well-known consequence of such a complementarity between substance and process observables is the "quantum Zeno effect" (Sudarshan 1983) which can be highlighted in the model of bistable perception (Necker cube) previously presented: the Hamiltonian evolution of a quantum system (represented by a process observable) can be slowed -and even frozen - by observing it frequently (that is, by a frequent application of a substance observable on its state).

Let us now consider the possible application of the notion of complementarity in the psychosomatic domain. As mentioned in Chap. 2, the idea of complementarity of mental and bodily descriptions of the individual can be traced back to Spinoza's double-aspect monism. We mentioned Spinoza's line of thought according to which Extension and Mind are regarded as two intelligible and non-interacting aspects of the neutral substance, which is God or Nature. Spinoza's parallelism of aspects has been later re-interpreted by Russell in epistemic terms and, more explicitly, in agreement with Bohr's project mentioned above, it has been interpreted as a *complementarity of mental and physical descriptions* by Pauli (see Chap. 2). The latter application of the notion of complementary to the mind-matter connection has also been explored in detail by Primas, with a special focus on the role of time (Primas 2003): McTaggart's physical time, B-time, which parametrizes material processes, and mental time, A-time, relative to our inner experience of time, are regarded as *complementary representations* of time (see Sect. 2.4).

In addition to the application of the generalized notion of complementarity of descriptions to the mind-body connection, the second sense of complementarity, which refers to the non-commutativity of observables, has also proven very fruitful for tackling the mind-body problem on experimental grounds (Uzan 2014a, b). The reason is that, as will be explained hereafter in more detail, the possibility of defining couples of non-commutative observables on each of the mental and the somatic domains can be a sufficient condition of the existence of non-local correlations between them. For explaining this essential point, we first have to explore the possibility of defining such couples of complementary observables. If the complementarity of mental observables has already been successfully explored in the domain of cognition and decision making (see above), it seems that it is not yet the case for observables relative to the *subjective experience,* that is, observables that evaluate the *emotional content* of a subject's mental states. Likewise, it seems that the complementarity of somatic observables, or at least of *physiological observables,* which are defined from the macroscopic physiological quantities that can easily be measured (such as heart rate, blood pressure or skin conductance), has not yet been explored.

4.2.2 Complementary Emotional Observables

Emotional observables measure the intensity of specified emotions experienced by a subject, such as fear, anger, sadness, disgust or happiness. These intensities can be collected by asking subjects to report discrete values on a graduate scale or to report them continuously, using a continuous response digital interface (CRDI) on which the subject moves a stylus or finger (Geringer et al. 2004). Let us concentrate here on the definition of emotional observables and their complementarity, which can be shown by computing their commutator. Emotional observables can be defined if we can compute from these data the conditional probabilities that an individual be "observed" (through a questionnaire or other means) in an emotional state given that she has been conditioned in a specified emotional state by a reliable method. This can be done from series of relevant data provided in several papers, such as the one of Prkachin et al. (1999) we will now refer to.

These authors have reported data on the measurement of cardiovascular variables on subjects experiencing emotional states induced by particular stimuli designed according to Lang's theory (Lang 1979). These stimuli are capable of triggering relatively "pure" emotions in subjects by referring to specific incidents in their lives. In the present case, the experimenter tells the subjects stories that are directly related to these specific incidents, but other methods based on the projection of films have also been shown reliable (Kassam and Mendes 2013). The five considered emotional contents are here "happiness" (noted as H), "sadness" (S), "anger" (A), "fear" (F) and "disgust" (D). In order to show the reliability of the method of conditioning, Prkachin and his colleagues have measured the average reported intensity (average on a population of 31 individuals, 23 females and 8 males) of the five basic emotions experienced in the six possible trials (the five trials of induced specific emotions and a trial where no specific emotion is induced). For our purpose, we shall here exclusively focus on these data. The intensity of the experienced emotions is measured on a scale of 0 (no emotion experienced) to 7 (extremely intense emotional experience). The results, which are presented in Prkachin's paper (p. 259), are reported hereafter (Table 4.1).

Table 4.1 Average intensity of each of five emotions experienced on six trials

	Emotional rating				
Target	Happiness	Anger	Fear	Sadness	Disgust
Neu	0.24[a]	0.03	0.09	0.15	0.03
	(0.66)	(0.17)	(0.38)	(0.57)	(0.17)
Hap	**5.30**	0.03	0.24	0.21[a]	0.15
	(1.36)	(0.17)	(0.71)	(0.49)	(0.87)
Ang	0.00	**5.15**	0.39	0.81[s]	1.21[a]
	(0.00)	(1.15)	(1.12)	(1.36)	(1.62)
Fea	0.21	0.73[a]	**4.61**	1.09[a]	0.36
	(0.74)	(0.31)	(1.50)	(1.59)	(1.08)
Sad	0.06	1.27[a]	0.79[a]	**5.15**	0.36
	(0.24)	(1.81)	(1.50)	(1.62)	(1.06)
Dis	0.06	1.58[a]	0.58[a]	0.85[a]	**4.97**
	(0.24)	(1.94)	(0.97)	(1.46)	(1.40)

From Prkachin et al. (1999)

For example, we can read on the fourth line (target emotion = fear), that a subject submitted to a fear-induced stimulus will, on average, report an intensity of fear of 4.61 (with a standard deviation SD of 1.50) and an intensity of happiness of 0.21 (with SD = 0.31). All the data of this table show that among the reported emotions the target emotion, which is induced by the trial, obtains a much higher rating than all other emotions, which shows the reliability of the method. Nevertheless, note that the target emotion is never reported with its maximal intensity (that is, with a rating of 7), which shows that the emotional state of a subject is not totally determined by the conditioning. We will then write a state of induced emotion B as W_B to distinguish this state from the state of "pure" emotion B (which would give rise to a rating of 7).

Now, for each reported emotion, these data do not provide the numbers of subjects that respectively report each of the 8 possible ratings, from 0 to 7, but a single number for its average intensity on the surveyed population. We cannot then work in an 8-dimensional state space corresponding to the 8 possible weightings of each reported emotion, as is usually the case for the quantum models of cognition and decision making (Busemeyer and Bruza 2012). Nevertheless, the average intensity of each reported emotion can be used to estimate the conditional probability to report this emotion for a subject prepared in a given state of induced emotion. From Table 4.1, the conditional probability of reporting emotion A for a subject prepared in a state of induced emotion W_B will be defined as the rate of the reported average intensity of A with respect to the sum of all the reported average intensities of emotions for subjects prepared in W_B. For example, the probability to report an emotion of disgust D for a subject prepared in an emotional state of anger W_A can be computed from the third line of Table 4.1 as $p\ (D/W_A) = 1.21 / 7.56 = 0.160$.

In order to show the complementarity of emotional observables, it is convenient to refer to a geometrical representation. The emotional states will be represented within a 2-dimensional state space where each emotion E defines the orthonormal E-basis $\{|E>, |\urcorner E>\}$, the vector $|E>$ being the state of extreme emotion E (with a rating of 7) and the vector $|\urcorner E>$ the state where no emotion E is felt (rating 0). This 2-dimensional state space is a complex Hilbert space capable of representing incompatible emotions (see hereafter).[8] In this state space, an emotional state is represented by a unit vector, while emotional observables are defined as projectors acting on this two-dimensional state space. For example, the Fear observable and the Sadness observable are respectively defined by the projectors $\mathbf{F}_{op} = |F><F|$ and $\mathbf{S}_{op} = |S><S|$. The probability *amplitude* that a subject reports the emotion S if they have been conditioned in a state W_F of induced fear can be computed from Table 4.1 by taking the square root of the corresponding conditional probability. This probability amplitude is given by the magnitude of the projection of $|W_F>$ onto $|S>$: $<S|W_F> =_{df} (P\ (S / W_F))^{1/2} \equiv$ length OH.

[8] Note that this geometrical representation of incompatible emotions is quite similar to that of incompatible spin observables in quantum mechanics, in a two-dimensional state space.

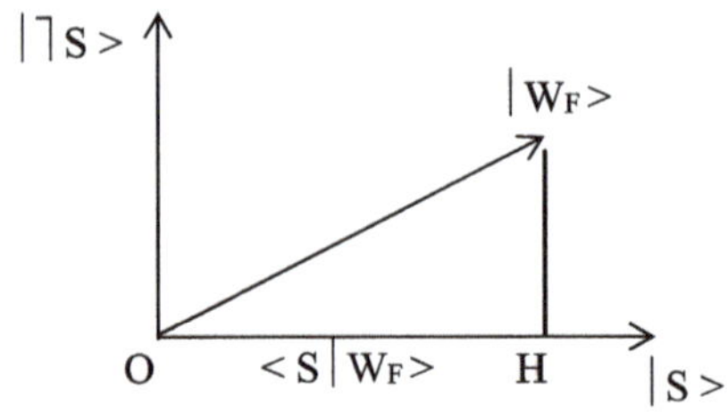

To show the complementarity of two emotional observables, we have to show that their commutator is different from zero. For example, let us consider the emotions Anger (A) and Disgust (D), whose associated projectors are, respectively, the projectors $\mathbf{A}_{op} = |A><A|$ and $\mathbf{D}_{op} = |D><D|$. The following diagram shows in the same 2-dimensional emotional state space the basis {|A>, |˥ A>} and {|D>, |˥ D>} respectively associated to these emotions, as well as the angles between these vectors and the state of induced anger $|W_A>$ that must be used to compute the components of the vector |D > in the basis {|A>, |˥ A>}:

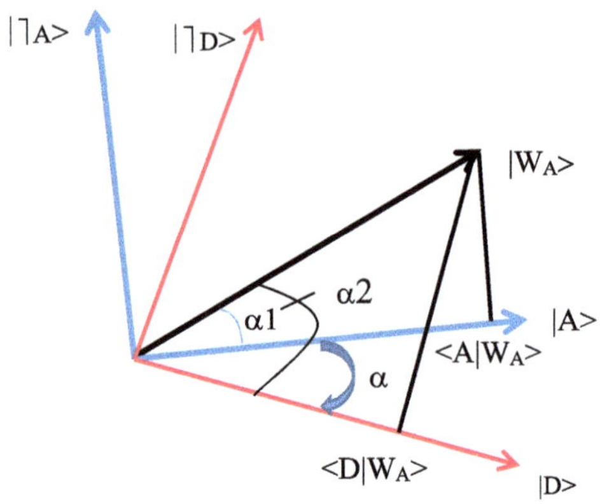

The calculation of the commutator [$\mathbf{A}_{op}$, $\mathbf{D}_{op}$] from the geometrical considerations on this diagram is given in detail in Appendix 2. It shows the *complementarity of the observables* $\boldsymbol{A}_{op}$ and $\boldsymbol{D}_{op}$, their degree of incompatibility being estimated by the norm of this commutator:

$$\|\left[\mathbf{A}_{op},\mathbf{D}_{op}\right]\| = 0.449.$$

Indeed, it can be shown that all couples of emotions involved in these data are complementary even if for some couples, such as happiness and anger, this commutator is very small. This means that, as expected, emotions cannot be considered as mutually exclusive. Many other examples of complementary emotional observables can be provided from the relevant experimental data reported in the literature, such as the papers by Kassam and Mendes (2013), Pauls and Stemmler (2003) and by Sinha et al. (1992) on which the meta-analysis of Sect. 4.4 is based.

4.2.3 Complementary Physiological Observables

In the physiological domain, it seems that no data about successive measurements of quantities has been systematically reported. We will then use a theoretical method to define couples of complementary physiological observables by generalizing the one which is used for defining conjugate observables in standard quantum theory (like position and momentum). The underlying idea is that conjugate *variables* of classical physics (like pressure and volume, pulsation and time,...) can be used to define complementary physiological *observables* because, as can be shown, their measurements are *interdependent*. Let us first give an example of such conjugate variables and show how their associated operators can be defined. To say that the variables p (pressure) and v (volume) are conjugate variables in the thermodynamic description of a fluid means that they are always associated to each other, as respectively a generalized "force" (intensive variable) and the conjugate generalized "displacement" (extensive variable), when calculating the change of the internal energy of this system – and this is also the case for the couples temperature/entropy and chemical potential/number of molecules:

$$dU = -p\ dv + Tds + \Sigma_i \mu_i dn_i + \ldots$$

Now, the couple of conjugate variables p, v can be used for defining *complementary cardiovascular observables* in as much as they are involved in the definitions of the blood pressure and the stroke volume, which is the volume of blood pumped by the heart in a cycle. As shown by several relevant physiological studies, these cardiovascular variables can be easily measured. The interdependency of the measurements of the blood pressure and the stroke volume in a cardiac cycle can be described as follows: To measure the volume of blood in a ventricle of the heart, we send ultrasounds on it (echography), which exert a pressure on this cavity and then change the blood pressure. Reciprocally, to measure the blood pressure, we change the blood flow by exerting a pressure on the arm's artery (by using a cuff filled with air and progressively emptied), which changes the volume of blood pumped by the heart.

This idea of interdependency in the measurement of physiological quantities can be implemented by defining complementary physiological observables acting on the physiological state space. Due to the fact that physiological quantities are continuous, this state space is here an infinite-dimensional complex Hilbert space H spanned by the vectors associated to the possible values (which are supposed non-degenerated) of all physiological quantities. The physiological state of a subject is represented by a unit vector of the state space H.

To implement the previous example, let us first define a *pressure observable*, which computes the blood pressure of the left ventricle of the heart of a subject in state |W > of H. This observable can be defined as:

$$\langle p \mid \mathbf{P}_{op} \mid W \rangle =_{df} p \langle p \mid W \rangle$$

in the "p-representation", that is, in the continuous basis {|p>} of the eigenvectors of $\mathbf{P}_{op}$ associated with its possible values. A straightforward correspondence can be established between the expectation value of the pressure observable $\mathbf{P}_{op}$ and the systolic blood pressure (SBP), which is the maximal blood pressure that can be measured:

$$\mathrm{SBP} = \langle \mathbf{P}_{op} \rangle_{ED}$$

where ED, the end of diastole state, specifies the moment of the cardiac cycle when the blood pressure is at its maximum value.

Let us now consider a volume observable, defined as:

$$\langle v \mid \Delta \mathbf{V}_{op} \mid W \rangle =_{df} (v - v_{ES}) \langle v \mid W \rangle,$$

where v_{ES} is the end-systolic volume of the left ventricle of the heart (which is the minimal volume of blood in this ventricle). This observable measures the increasing in the volume of blood contained in this ventricle of the heart with respect to its minimal value v_{ES}. The interest of this definition is that the stroke volume (SV), which is the difference ($v_{ED} - v_{ES}$), is equal to the expectation value $<\mathbf{\Delta V}_{op}>$ of the observable $\mathbf{\Delta V}_{op}$ at the end of diastole:

$$\mathrm{SV} = \langle \Delta \mathbf{V}_{op} \rangle_{ED} \, .$$

To show the complementarity of $\mathbf{P}_{op}$ and $\mathbf{\Delta V}_{op}$, the pressure and volume observables, we have first to show that the p-representation <p |W > and the v-representation <v |W > of any physiological state |W > are Fourier transforms. This can be done by noting that <p |W > and < v |W > are square-summable functions that admit (by Plancherel theorem) a Fourier decomposition. The phase of the harmonic functions involved in these Fourier decompositions can be estimated, for a couple of blood pressure p and an additional volume Δv of the left ventricle of the heart, as the rate of p. Δv to its maximal value ΔU. The latter quantity ΔU is the mechanical work of the heart in a cycle when the temperature remains constant and no chemical reactions take place (which is indeed the case here). Due to homeostasis, ΔU can be considered as a constant for a given subject (Balaban 2012) and, taking into account that the right ventricle pressure is about seven times smaller than the systolic pressure SBP, it can be calculated as (Uehara and Koibuchi 2011):

$$\Delta U = 1.14\,\mathrm{SBP} \times \mathrm{SV}.$$

By writing now the operators $\mathbf{P}_{op}$ and $\mathbf{\Delta V}_{op}$ as Fourier transforms, and by computing the action of the observable $\mathbf{P}_{op}$ in the v-representation, we can establish that $\mathbf{P}_{op}$ is the following differential operator:

$$\langle \mathrm{v} \,|\, \mathbf{P}_{op} \,|\, \mathrm{W} \rangle = -\mathrm{i}\,\varepsilon\,\partial / \partial \mathrm{v} \langle \mathrm{v} \,|\, \mathrm{W} \rangle,$$

and then their commutator is different from zero:

$$\left[\Delta \mathbf{V}_{op}, \mathbf{P}_{op} \right] = \mathrm{i}\,\varepsilon\,\mathbf{Id},$$

their degree of incompatibility ε characterizing the cardiovascular cycle of the subject under consideration. This calculation, drawn from the reference (Uzan 2016), is explained in more detail in Appendix 3.

The value of $\varepsilon = \Delta U/1.14$, where ΔU is the (constant) maximal mechanical work of the heart in a cycle, can be estimated as follows from Malanga's work (Malanga 2007): the maximal pressure of the right ventricle of the heart is of the order of 8 mmHg ≅ 1066 Pa, while the maximum stroke volume for the right ventricle is about 70 ml = $7 \cdot 10^{-5}$ m^3, then $\Delta U \cong 7 \cdot 10^{-2}$ Joules and $\varepsilon \cong 6.54 \cdot 10^{-2}$. The norm of the commutator of the couple of observables {stroke volume $\mathbf{\Delta V}_{op}$; systolic pressure $\mathbf{P}_{op}$,} informs us about their degree of incompatibility:

$$\left\| \left[\Delta \mathbf{V}_{op}, \mathbf{P}_{op} \right] \right\| \cong 6.54 \cdot 10^{-2} \text{ Joules}.$$

This method can of course be applied to other couples of conjugate physical variables that refer to complementary descriptions of the same physiological phenomenon, as is the case for the temporal and the spectral descriptions of the heartbeat. In this case, we can define complementary observables that respectively measure the duration Δt of a heartbeat and its spectral dispersion Δf. These two observables, respectively a time observable $\mathbf{T}_{op}$ and a frequency observable $\mathbf{F}_{op}$, can be defined as follows:

$$\langle \Delta t | \mathrm{T}_{op} | \mathrm{W} \rangle =_{df} \Delta t \langle \Delta t | \mathrm{W} \rangle, \text{in the continuous duration} \{ |\Delta t > \} \text{representation},$$

$$\langle \mathrm{v} | \mathrm{F}_{op} | \mathrm{W} \rangle =_{df} \Delta f \langle \Delta f | \mathrm{W} \rangle, \text{in the continuous spectral} \{ |\Delta f > \} \text{representation}.$$

$\mathbf{T}_{op}$ and $\mathbf{F}_{op}$ are complementary observables since the duration representation and the spectral representation are Fourier transforms. The latter property is due to the fact that Δt and Δf are *dual* quantities that satisfy the following relation (according to which the greater the duration of the signal, the smaller the spectral dispersion):

$$\Delta f\, \Delta t = \mathrm{C} / 4\pi,$$

where the constant $C \geq 1$ characterizes the form of the considered wave packet ($C = 1$ only for a Gaussian wave packet). The commutator of the time observable and the frequency observable can then be computed as $[T_{op}, F_{op}] = i\,(C/4\pi)\,Id$, whose norm is: $\| [T_{op}, F_{op}] \| = C^2 / 4\pi^2 \geq 6.3 \cdot 10^{-3}$.

The degree of incompatibility of these conjugate observables can be very small in comparison to the one of emotional observables (see above). This means that there is only a few indeterminism of the values of the physiological observables, which has led to the wide-spreading opinion that these observables are nothing but pure classical observables whose values are perfectly determined at each moment, independently of any measurement. However, despite of these small values, a fundamental uncertainty on somatic observables exists and, combined with that of the emotional observables, it gives rise to *non-local* mind-body correlations, in accordance with what was asserted in Sect. 3.4.

The previous analysis shows that the notion of complementarity can be implemented in psychophysics, both for the subjective experience and the physiological description of the individual. The emotional, complementary observables have been defined by referring to experimental data on measurements of experienced emotions reported by subjects prepared in a given state of induced emotion. In the physiological domain, complementary observables corresponding to cardiovascular quantities have been defined by generalizing the notion of conjugate observables in quantum mechanics. The few examples presented here can be easily completed by considering several couples of psychological and physiological observables defined from other independent data. As will be explained below, the possibility of defining couples of complementary psychological observables and complementary physiological observables paves the way for an experimental test of the nature of the mind-body connection, and in particular of the suggested interpretation in terms of generalized entanglement.

4.3 Entanglement in Quantum Physics and Beyond

In the quantum domain, the notion of "entanglement" was originally introduced by Schrödinger to characterize the impossibility of describing two quantum systems which interacted in their common past by two independent states (Schrödinger 1935, p. 555). The notion of entanglement captures the idea of inseparability of the descriptions of these two sub-systems. A physical system S composed of two parts S_I and S_{II} is an entangled state if this state is *non-factorable,* meaning that it cannot be written as the product of a single state of S_I and a single state of S_{II}. Let's call this vector-state $|\Phi>$. Its general form in the tensor-product space of the respective state spaces of S_I and S_{II} is (see above):

$$|\Phi>= c_1 |\varphi^I{}_1>|\varphi^{II}{}_1>+c_2 |\varphi^I{}_2>|\varphi^{II}{}_2>+\ldots..,$$

where the products $|\varphi^I{}_i>|\varphi^{II}{}_i>$ are the possible global states of S. If $|\Phi>$ is entangled there is no pair of vector-states $|\Phi_I>$ and $|\Phi_{II}>$ of S_I and S_{II}, respectively, such that $|\Phi> = |\Phi_I> \otimes |\Phi_{II}>$. In this case, there are correlations between S_I and S_{II} *even if S_I and S_{II} are causally separated*[9]: if S_I is found in the state $|\varphi^I_i>$, S_{II} will necessary be found in the corresponding (or "relative") state $|\varphi^{II}_i>$, whatever their distance and without the need of any exchange of signal between them. Moreover these a-causal correlations cannot be explained classically, by the pre-existence of intrinsic values of the measured observables (property of non-locality). In particular, the existence of these correlations cannot be assigned to a common cause.

This phenomenon has been successfully tested in many experiments, such as in Aspect's experiments (Aspect 1982), by measuring the polarization of two photons prepared in such an entangled state:

$$|\Phi> = \frac{1}{\sqrt{2}}\left(|\uparrow>|\downarrow> + |\rightarrow>|\leftarrow>\right),$$

where $|\uparrow>$, $|\downarrow>$ and $|\rightarrow>$, $|\leftarrow>$ respectively denote the possible vertical and horizontal polarization states of the photons (Fig. 4.3). These photons are emitted in two opposite directions and their polarization is measured according to different angles by the appropriate devices D1 and D2:

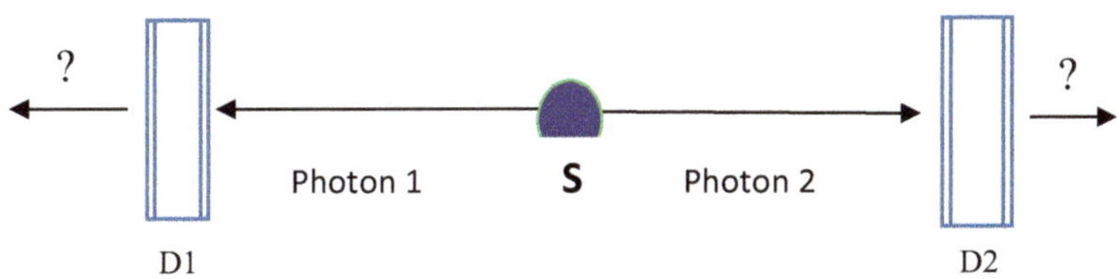

Fig. 4.3 A schematic diagram of Aspect's experiment

The outcomes of measurements performed on a big number of such "twin photons" show that their polarizations are strongly correlated, in agreement with the prediction of quantum theory. Yet, the experiment is performed in such a way that any possible exchange of information between them or between the different parts of the experimental device has been ruled out.[10] Moreover, what is essential, the particles did not actually possess *before the observation* the properties (polarisations) measured by the devices D1 and D2. This fundamental point has been established by the violation of some statistical conditions known as Bell's inequalities.[11]

[9] That is, there is no possible causal interaction between these two sub-systems.

[10] In the words of relativity, the measurement-events of these properties (spin) are said to be spatially separated.

[11] Bell's inequalities are conditions that must be satisfied for any statistics bearing on a population whose individuals are supposed to intrinsically carry their properties. The first formulation of such conditions was established by the physicist John Bell (1964), but Aspect's experiment tested an equivalent formulation, more suited to this experimental test, which was proposed by Clauser, Horne, Shimony and Holt (1969), called CHSH inequalities. The CHSH inequalities, which capture the classical limit of Landau's inequalities, when all observables are commuting, are displayed in the next paragraph on generalized entanglement.

These correlations cannot thus be explained classically, neither as causal interactions nor by the existence of the sub-systems' predetermined intrinsic properties.

4.3.1 Generalized Entanglement

Similar to the concept of complementarity that has been defined and explored above, the notion of entanglement can be generalized beyond physics. Such a generalization relies on the possibility of applying the quantum *formalism* beyond the physical realm. An entangled state of a generalized system will thus be written as *a non-factorable state* of the tensor-product of the Hilbert spaces respectively associated to its subsystems. As mentioned above, this representation has already been used for dealing with entangled cognitive states (Aerts et al. 2011; Busemeyer and Bruza 2012). Mind-body entanglement refers to the a-causal and non-local correlations between the mental and the somatic aspects of a person, regarded as an indivisible unity. The existence of such correlations is closely linked with the possibility of defining complementary mental observables and complementary somatic observables. This link has been shown by Landau (1987) within the C* algebra of observables that underlies the standard quantum theory. Landau's theorem is a generalization of the CHSH inequalities, which are used for showing the (non-) locality of correlations, like in the previously mentioned Aspect's experiments. Let us briefly mention this fundamental result and its extension to the general systems theory we work in here.

Let us consider two couples of observables A, B and A', B′ respectively measured on two *causally separated* regions, called S_I and S_{II}, of a system S (Fig. 4.4). Landau's theorem sets out, in both the classical case (where all observables commute) *and* in the quantum case (of non-commutative observables), the conditions fulfilled by the correlation factor R, defined as:

$$(\mathrm{Df})\mathrm{R} = \mathrm{S}(\mathrm{aa}') + \mathrm{S}(\mathrm{ab}') + \mathrm{S}(\mathrm{ba}') - \mathrm{S}(\mathrm{bb}'),$$

where S (aa′) is the expectation value of the product of the results of measurement of A and A′, and we have a similar definition for the other terms.

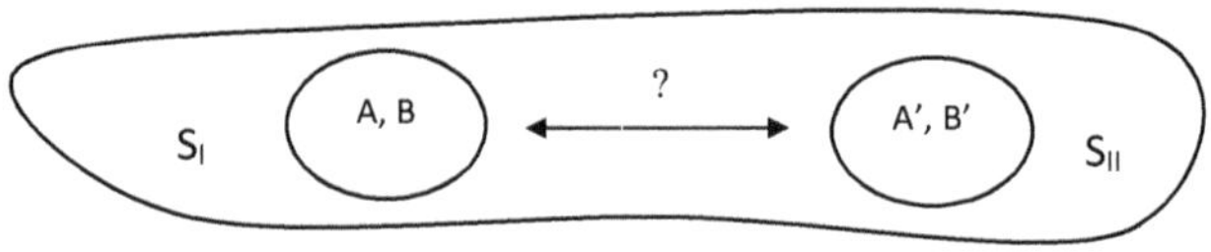

Fig. 4.4 How correlated are S_I and S_{II}?

As expected, this theorem reduces to the CHSH inequalities in the classical case of commutative observables:

$$\text{CHSH} \quad \text{For}\, S_I \text{ and } S_{II} \text{ separated, } [A,B] = 0 \text{ or } [A',B'] = 0 \Rightarrow -2 \le R \le +2,$$

which characterizes (classical) local correlations. However, in the quantum case of non-commutative observables, Landau's theorem provides us with a very powerful means to capture the essential link between the complementarity of observables defined on the two sub-systems and the non-locality of their correlations. This theorem shows that *the possibility of defining non commutative properties on each of the two causally separated parts of a system can be a sufficient condition for ensuring the existence of non-local correlations between these two subsystems (which is expressed by the fact that $|R|$ is bigger than 2, the classical limit)*:

(T) For S_I and S_{II} separated, $[A, B] \neq 0$ <u>and</u> $[A', B'] \neq 0 \Rightarrow -2\sqrt{2} \leq R \leq +2\sqrt{2}$.

We say that the condition $[A, B] \neq 0$ and $[A', B'] \neq 0$ "can" be a sufficient condition for ensuring the existence of non-local correlations between the two subsystems because the double-inequality $-2\sqrt{2} \leq R \leq +2\sqrt{2}$, which can be more concisely written as $|R| \leq 2\sqrt{2}$, actually describes *two* possibilities: $|R| \leq 2$ or $2 > |R| \leq +2\sqrt{2}$. The non-locality of these correlations is thus established only in the second case, when $|R| > 2$.

In this theorem (T), the number "$2\sqrt{2}$", called Tsirelson bound from the name of the mathematician who first established it, provides the maximal amount of correlations between the two subsystems (Tsirelson 1980; Landau 1987). Note that the Tsirelson bound can also be established from purely informational considerations (Pawlowski et al. 2010; Bub 2012), which shows that it is in fact a very general no-signaling bound, which is independent of the quantum formalism (Uzan 2018, 2020).

Even more precisely, Landau (2001) has shown that for separated systems and binary (+1/−1) observables the quantity $<c^2>_{|\phi>}$, which denotes the expectation value of c^2 in the state $|\phi>$, c being defined by $R = <c>_{|\phi>}$, can be computed as follows (Landau 2001, Sect. 8):

$$\left\langle c^2 \right\rangle_{\phi>0} = 4 + \left\langle \; A,A' \quad B',B \; \right\rangle_{\phi>0}$$

Consequently, by the property that the absolute expectation value of any observable in a state which is the eigenvector associated to its maximum (supposed non-degenerated) eigenvalue is equal to its norm, the maximum of the expectation value involved in the right member of the previous equality is equal to the norm $\| [A, A'] [B', B] \|$. Then, under the condition that S1 and S2 are separated, and for any state $|\phi>$ of the composed system, the maximum value of the right member of the last inequality is $\beta = (4 + \| [A, A'] [B', B] \|)^{1/2}$. This result can be synthesized as follows[12]:

> (Th) Two *causally separated systems* on which are respectively defined the couples of observables (A, A′) and (B, B′) satisfy the condition $|R| \leq \beta$, where $\beta = (4 + \| [A, A'] [B', B] \|)^{1/2}$.[13]

[12] This more general result of course contains the CHSH inequalities by imposing that all the couples of observables are commuting (classical case) and Tsirelson inequalities (T) by assuming that the two commutators are different from zero.

[13] An order of magnitude of the signalling bound β for the two couples of complementary observables (Anger, Disgust) and (Systolic pressure, Stoke volume) considered above is 2.1 is obtained by using those of the relevant commutators $[A_{op}, D_{op}]$ and $[\Delta V_{op}, P_{op}]$ calculated above and the definition of β.

Note that this theorem still holds under the weaker assumption that the *mean values* < A>, <A′>, <B > and < B′ > of the observables are bounded by 1 –for, as explained in the reference (Uzan 2014a), all the steps of its derivation remain valid under this weaker assumption. This remark will be very useful in the next Sect. 4.4 for analyzing available experimental data.

A careful study of the derivation of these results seems to tell us that they are very general since it *does not a priori refer to any physical properties of the considered "systems"*.[14] The applicability of these results within the presented general systems theory obtained by generalizing quantum theory has been established rigorously (Uzan 2014a; 2020). It has first been shown that they can be applied to *generalized systems* defined and characterized as above; second, it has been shown that they remain valid for *continuous* observables whose spectrum is bounded by 1; and third, the converse of these conditions has been demonstrated by prolonging to generalized systems and continuous observables Gisin's derivation of the converse of Bell's theorem (Gisin 1991) and Landau's theorem previously mentioned (Landau 1987), which allows a more accurate characterization of the exchange of signals, whatever they are, between the subsystems. To sum up, the generalized systems theory based on the key-quantum concepts presented above derives the following results that characterizes the nature of the correlations between the two parts of a generalized system:

(GTh) Let S_I and S_{II} be two parts of a composite general system on which are respectively defined the couples of (possibly) continuous, bounded observables (A, A′) and (B, B′), and let R be the CHSH correlation degree between S_I and S_{II}. The following statements are then verified:

1. $|R| \leq 2$ if and only if these correlations are due to intrinsic, predetermined values of the observables (case of a-causal *classical* correlations),
2. $2 < |R| \leq \beta$ if and only if these correlations are *a-causal* and *non-local* (case of entanglement correlations).
3. $\beta < |R|$ if and only if these correlations are due to an exchange of signal (case of causal interactions).

These results are displayed on the following diagram:

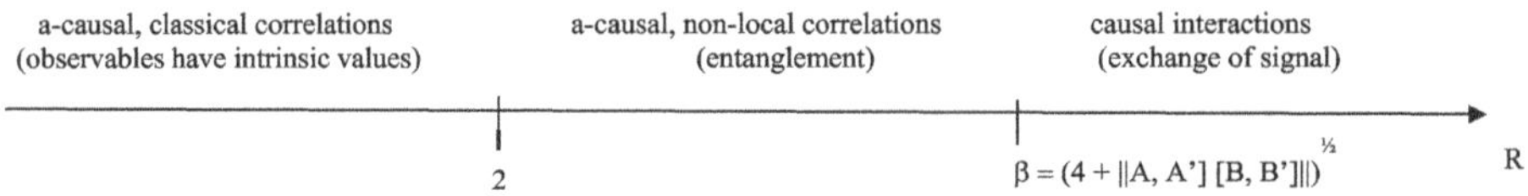

[14] They are established within the non-commutative algebra of operators that underlies the standard quantum theory and does not thus involve any a priori reference to the physical world (no Planck constant, no energy or mass concepts, etc.…) but only operators, which are purely mathematical entities, endowed with a structure of C*-algebra. This demonstration also uses Schwartz inequality applied to expectation values, which asserts that $|E(X)|^2 \leq E(X^2)$, where X is a random variable and E(X) its expectation value. See the references (Landau 1987; Uzan 2014a, 2020).

Now, these results will be employed to determine *the nature of the correlations between mental and somatic observables.* This question will be decided by comparing the CHSH correlation factor defined from couples of mental and physiological observables with the Bell bound and the signaling bound β relative to these observables. For this, we have to follow these steps:

1. The subsystems S_I and S_{II} are to be thought of as the mental and the somatic aspects of the individual, respectively.
2. Couples of complementary mental observables and complementary somatic observables are to be defined on S_I and S_{II}, from the relevant experimental data and from theoretical considerations on conjugate variables, respectively.
3. Taking now into consideration theorem (GTh) above, one can determine the nature of the psychophysical correlations considered from the values of the relevant CHSH correlation factor.
4. In particular, if this correlation factor, noted as R, significantly satisfies the double inequality $2 < |R| \leq \beta$, the considered mind-body correlations are relations of generalized entanglement, that is, they are a-causal and non-local correlations which cannot be interpreted as interactions or totally explained by the existence of intrinsically defined mental and somatic properties.

Note that this experimental test can also be used to show the existence of *temporal non-locality*, which we alluded to in the conceptual presentation of the notion of mind-body entanglement in Chap. 3 (Sect. 3.4). This means that the couples of observables (A, A′) and (B, B′) can be respectively measured *at distinct times t and t′ on the same system.* In this case, the correlations between these two couples of observables cannot be explained causally, referring for example to the temporal evolution of the system under consideration. This rather counter-intuitive consequence of the mind-body entanglement that regards temporal non-locality plays an essential role in explaining the notion of "body memory" and its use for healing. This point will be explored in Chap. 5.

4.4 An Experimental Test of Mind-Body Entanglement

As explained above, it is possible to provide an experimental response to the question of the nature of the mind-body connection by evaluating the correlation degree between mental and somatic observables. Beyond the metaphysically-based options presented in Chap. 1, this response relies on the scientific method, appealing to theoretical models and hypotheses *that can be tested experimentally.*

In order to implement this experimental approach to the mind-body problem, a method is to measure physiological variables on subjects conditioned in specific emotional states. For example, we can measure cardiovascular variables on subjects experiencing emotional states induced by telling them stories that are directly related to specific incidents in their life, following Lang's method (Lang 1979), or by showing them videos, like in the experiments reported below, in order to induce a specific emotional state. Another idea is to measure the heart rate variability (HRV) of subjects in specific emotional states. Some ECG measuring device can extract the changes in heart rate frequency from a classical electrocardiogram and make its spectral analysis. This device can compute the ratio of very low, low and high frequency of heart beat, and then indicate the amount of sympathetic and parasympathetic activities. These antagonist activities can be used to define complementary, physiological observables that will be measured alternatively while the emotional observables evaluate for example their degree of mental relaxation, animation or arousal by using questionnaires. Other physiological observables, like those related to the brain electrical activity (EEG) or the skin conductance, can also been measured conjointly with emotional observables. Once normalized (values between -1 and $+1$), the measured couples of mental-physiological values can be used to compute the CHSH correlation factor.

Preliminary results of this statistical test of the nature of psychophysical correlations has been obtained by carrying out a meta-analysis of independent data reporting the change of physiological variables on subjects reliably conditioned in specific emotional states. The reference (Uzan 2017) presents such a meta-analysis of the experimental data provided by the articles of Kassam and Mendes (2013), Kreibig et al. (2007), Sinha et al. (1992) and Prkachin et al. (1999). These articles provide a diversity of data relative to the joint measurements of emotional and cardiovascular variables. The raw data are presented without a priori interpretation of the measured correlations, for example by computing right away Pearson's or Spearman's correlation factors, which is of no interest here. For each of the trials, that is, for each set of couples of emotional observables (A, A′) and cardiovascular observables (B, B′), the CHSH correlation factor R has been computed and its magnitude has been compared to the Bell bound and (if applicable) to the signaling bound β, which is computed from the commutators of the considered observables. In order to not make this text too heavy, we are not going to present here all the calculations of this meta-analysis (normalizations of data, computation of R, computation of the signalling bound, and statistical considerations), as this is presented in detail in the reference (Uzan 2017). However, Appendix 4 describes the main steps of this meta-analysis and illustrates them by referring to some of the considered experimental data.

4.4.1 Conclusion of this Meta-Analysis Covering 168 Subjects and Discussion

For the 168 subjects taking part to this meta-analysis, the CHSH correlation factor R between all the available couples of emotional and cardiovascular variables and for the protocol described above (which, in particular, regards emotional variables as *categorical*) has been distributed as follows:

1. Regarding the question of psychophysiological causation: 142 values satisfy the Bell bound *and then the signalling bound* ($|R| \leq 2 < \beta$), which represent a proportion of 84.52%, while 15.47% violate the Bell bound. This shows a large

advance to the correlations that satisfy the signalling bound. Assuming that the distribution of the values of R roughly respects a normal distribution whose standard deviation can be approximately calculated (by the binomial law) as SD = $(168 \times 0.8452 \times 0.1547)^{1/2} = 4.68$, we can assert that, with a confidence level of 95% (corresponding to a deviation of the mean value within 2 SD = 9.36), R satisfies the signalling bound ($|R| \leq \beta$) with a probability comprised between (142 + 9.36)/168 = 90.1% and (142–9.36)/168 = 78.9%. Under the conditions of experiments that will be discussed hereafter and the precision of the reported data, it can then be concluded that a significant part of the psychophysical correlations that are involved in these experiments are ***non-signaling correlations*** with a good level of confidence. This means that the psychophysical correlations cannot be explained by the existence of an "interaction" or an exchange of signal. In other words, the reported experimental data show with a good level of confidence that the psychophysical correlations under consideration are constitutive of the individual and not due to an elusive and rather mysterious "interaction" between bodily and mental processes. This conclusion, if confirmed (see below), would show, in the extension of Darwin's evolutionist theory of emotions mentioned at Chap. 3, that the mind-body correlations have been established during the evolution process understood in a broad sense: in the phylogenetic evolution, as a member of the human kind; ontogenetic evolution, as the offspring of one's parents; and even socio-cultural evolution, though personal experience.

2. Regarding mind-body entanglement, this meta-analysis does not show significantly that $|R|$ is bigger than 2 and less than the relevant signalling bounds. The existence of non-local psychophysiological correlations is then not clearly established. Indeed, we did not even compute β for all the couples of emotional and cardiovascular observables because only a few values of R were greater than 2. If this experimental test is highly conclusive for establishing the a-causality of these correlations, a more accurate test must be realized to establish their non-locality. The evaluation of the correlation factor R could indeed be made much more reliable by *continuously* measuring the emotional variables instead of considering the induced emotions as maximally experienced, which surely skews the results since all the (realistic) situations where emotional observables do not have a maximal intensity have been ruled out of this analysis. Another issue of this statistical study is the use of mean values of cardiovascular observables over the population. This point is questionable since in the theorem (ThG) of Chap. 4 on which this statistical test is grounded, R is defined as a combination of expected values for the same subject's state. However, in spite of the reliability of the emotional conditioning, all the tested subjects are not supposed to be "prepared" in exactly the same psychophysical state and to have exactly the same bodily correlates to emotions. Two different subjects, even prepared in very similar emotional states, may trigger very different biological responses to these emotional conditioning depending of their own psychosomatic structure. In order to be more significant, this test should then select groups of subjects with similar sensibility to the method of emotional conditioning and similar reactivity to emotions. This means

that for these subjects the specific couples of observables under study should have a similar statistical distribution, which could be realised by choosing a group of individuals with very close cardiovascular characteristics and very close temperaments, which needs to refer to some known psychosomatic classification -see, for example, Janssen (2006, Chap. 5). Moreover, this statistical test should also consider other psychophysiological correlations by referring to other bodily observables -like, for example, the brain electrical activity or the concentration of some substances in blood. A further statistical test is planned with a larger number of test subjects with similar psychophysical features and by continuously tracking the emotional variables using a graphic interface (where the subject can move their finger along a ladder). If the case of signalling correlations seems to be reliably ruled out by the previous experimental results, it is expected that further results will show that for some statistical series the CHSH correlation factor significantly satisfies the condition of non-locality $2 < |R| \leq \beta$, at least for some couples of mental and somatic observables.

A final clarification must be made about this experimental test. In this meta-analysis, we have obtained significant results refuting the "signaling" interpretation of psychophysical correlations. Some readers could think that the signaling characterization of causation relies on a purely mechanical and spatial interpretation of causation, understood as an exchange of force, moment or energy, and that, consequently, only this mechanical and spatial interpretation has been tested experimentally. However, the notion of "signaling" is much more general since it refers to the possibility that one of the two domains (mental or physical) is *in-formed (literally, put in form) by what happens in the other.* "Signaling" thus generally refers to *an exchange of information,* which can be a priori independent of any material connection through space. The signaling interpretation of causation, which relies on the processual approach presented in Chap. 1, is indeed the *minimal requirement of the concept of causation* needed for distinguishing it from that of "explanation" provided within a given paradigm (see Sect. 1.1). What this meta-analysis shows is thus the impossibility of causally interpreting the mind-matter connection, regardless of any specific interpretation of causation that meets this minimal requirement. This conclusion corroborates the logico-philosophical analysis of Chap. 1: the concept of causation is *unsuitable* to interpret the mind-body connection.

Chapter 5
Mind-Body Entanglement and Healing

The biomedical model of health exclusively focuses on the body's dysfunctions, or even on the dysfunctioning of a single organ or a single genetic anomaly supposed to be causally responsible for the patient's illness. In contrast to this approach, and in accordance with Hippocrates' holistic line of thought,[1] the systemic, quantum-like approach presented here conceives of illness as an imbalance in the patient's psychosomatic state. Healing a person's illness will thus consist in rebalancing this state. This rebalancing can be done by changing the relative weights of the specific psychophysical correlations encoded in this state, which needs to take into account their subjective life in an essential way. Mind-body entanglement can explain the existence of a body memory which, in turn, explains the genesis and evolution of internal diseases and suggests the development of holistic therapies. In particular, the concept of mind-body entanglement provides new insights on mental disorders, their classification and their relationship with brain functioning. Furthermore, it provides a theoretical framework for explaining consistently paradigmatic self-healing techniques, which remain totally enigmatic within the classical, causal-local framework. This is the case of the placebo "effect", biofeedback and the healing power of meditation. Mind-body entanglement also explains the possibility of distant healing, for which significant data have been reported.

5.1 From Biomedicine to the Holistic Approach to Illness

The biomedical model of health developed over the course of the nineteenth century reduces the patient's illness to a disease, which is only its purely biological, "objective" aspect, regardless of the patient's suffering and of any social factor. In this

[1] See Sect. 3.1, which deals with the origins of the concept of psychosomatic unity.

P. Uzan, *Mind-Body Entanglement*, SpringerBriefs in Philosophy,
https://doi.org/10.1007/978-3-030-90173-8_5

ontological model, each symptomatic expression can be associated with organic damage and thus diseases are isolable, objectified entities that can be treated independently of the patient's subjective and social life (Laplantine 1993). Similar to the contamination of the biological body by an exogenous microbial agent and the successful model of vaccination, it proposes to heal the patient by only treating their body's dysfunctions through external interventions, recommending medication, surgery, and vaccine. For example, surgery is supposed to heal the patient by removing a tumor, thus excluding any psychological or social factor. According to the biomedical model, the role of medicine is reduced to understanding and to maintaining the body's functioning, as it conceives of it as a biological machine–a view inherited from Descartes' mechanical metaphor of the body he proposed in his Treatise on Man (Descartes 1664).

However, the biomedical model of health cannot be upheld because of the strong intertwinement of the biological, psychological and social aspects of a patient's illness. Nowadays, people are increasingly suffering from chronic disease, which do not result from exogenous aggressions and cannot be healed by only focusing on their biological aspect. A chronic or internal[2] disease is a long-term disease which develops progressively and significantly relies on endogenous somatic *and* mental factors developed throughout the patient story. This is, for example, the case of asthma, cancer or diabetes, whose origin seems to rely equally on genetic, behavioural, emotional and social factors (Trombini and Baldoni 2005). Even more clearly, in the psychiatric domain, long-term disorders like depression, schizophrenia or bipolar disorder cannot exclusively originate from biological factors. Trying to heal a patient's illness by only focusing on its biological aspect cannot succeed since essential information about their psychosomatic state is left out. As explained in Chap. 3, the person's psychosomatic state can neither be reduced to its somatic aspect nor to its psychic aspect because it contains an essential and irreducible information about their connection: it literally *encodes* their correlations (see Sect. 4.1). Focusing exclusively on the somatic description cannot then provide a complete information about the patient's illness. To be reliable, a healing process must take into account both mental and somatic aspects of a person's illness as well as their connection. Illness cannot be conceived of as an isolable and objectifiable entity of the body but it must, instead, be understood in terms of psychosomatic imbalance, in the spirit of the humoral medicine of Hippocrates where health was characterized by the balance of humors.

Let us mention that increasing number of doctors now integrate to their practice these other, non-purely biological elements of an illness in reaction to the reductionist biomedical approach. For example, it is generally recognized that a purely biomedical approach to heal a patient suffering from cancer is insufficient. As explained

[2] The two qualifiers "chronic" and "internal" can be used to name diseases which generally develop very gradually, are disabling and can last a very long time. "Chronic" indicates that their evolution is long and progressive, while "internal" rather underlines their endogenous origin. We will use the term "internal" throughout the rest of this book to emphasize the fact that they essentially rely on the patient's psychosomatic organisation.

by Professor Khayat (2013), what must be treated is not only the biological aspect of cancer, which is an anarchic cell multiplication that can pervade into the whole body, but also the patient's emotional experience (like their depressive state) and the social consequences of the situation (such as social distancing or threats concerning one's job). As explained in Chap. 3 from Broom's work, what the patient thinks and feels about their disease, as well as the meaning they assign to their symptoms is *a fundamental element of the healing process*.

Actually, the biomedical approach has been complemented by a more comprehensive approach centered on the whole person and not only on the biological aspects of their illness. In reaction to the shortcomings of the biomedical model, Engel developed a biopsychosocial model aiming to integrate these psychical and sociological elements into a more global approach to healing (Engel 1977). Engel's biopsychosocial model has been developed from the work of psychoanalyst and neurologist Grinker (1964) who referred to the biological paradigm of Bertalanffy's general systems theory (briefly presented in Sect. 3.4), going so far as to assert that biological reductionism is unscientific. This model is not only more suited to chronic or internal disease but, more fundamentally, it suggests to consider the whole person and their behaviours, conceiving of health and illness as resulting from "interactions" between biological, psychological and social factors. Correlatively, Engels' biopsychosocial model aims to counter the dehumanization of medicine and the infantilization of patients, which is inherent to the biomedical approach. It includes the patient's subjective experience into the healing process, emphasizing in particular the quality of the doctor-patient relationship which transforms the patient's role from a passive object of investigation into a subject, considered to be the protagonist of the medical act.

However, in spite of this important holistic turn, the biopsychosocial model still relies on a concept of causal "interactions", even if the latter are conceived as complex, non-linear causal processes. In particular, it takes for granted the existence of causal interactions between emotional and bodily aspects of illness. As was already the case for Bertalanffi's general systems theory, it still neglects the yet essential organizational principle making the psychosomatic unity possible, namely, mind-body entanglement. This means that the typical properties of the psychophysical correlations must be taken into consideration for completing and refining the biopsychosocial model of health. These properties of inseparability, meaningfulness (or specificity), a-causality and non-locality have been explained in Chap. 3 and clarified in Chap. 4 within a quantum-like framework. Taking now these essential features of the mind-body connection into account will enable us to set the basis of a novel holistic approach to illness, which is grounded on the concept of embodiment rather than that of "interaction".

5.2 Body Memory, Internal Diseases and Complementary Medicines

According to the systemic, quantum-like approach to the individual presented here, a person is an indivisible mind-body system whose mental and biological properties are correlated by generalized entanglement relations. As explained in Sect. 4.1, stating that a person's psychosomatic state is entangled means that this state must be written as a non-factorable weighted linear combination of elementary product-states of all possible bodily and emotional basic states. These weights, which are complex probability amplitudes, characterize a person's psychosomatic signature. Their emotional experience, described by the intensity of the basic emotions they feel, is, at any moment, strongly correlated with their body's state. There are two important consequences of this psychophysical entanglement. First, it is expected that a person's internal disease, which has been classified as purely "somatic", involving a set of physiological, endocrine and immune dysfunctions, is in close correspondence with some of their psychological conflicts. This view strongly suggests that healing one of these two aspects could have significant "effects" on the other, thus rebalancing the global psychosomatic state by increasing the probability of the associated product-states.[3] This direct way of healing can be illustrated by the placebo phenomenon, which shows that experiencing positive emotions induced by a person we trust can trigger body restorative processes, as well as by biofeedback, mental imagery and meditation that will be discussed and modelled in Sect. 5.3. These phenomena can thus be taken as *therapeutic resources* -which is presently not the case for the placebo phenomenon, which is only considered as a background noise to which the efficacy of a drug is compared. They define *a first direct mind-body mode of healing* that will be modelled hereafter within our systemic approach (Sect. 5.3).

However, this direct mode of healing does not always work in the case of repressed past emotional traumas, which prevents the subject from reaching such an appropriate mental state. In this case healing requires restoring or reorganizing the patient's body memory where these repressed emotional traumas have been imprinted. The notion of "body memory" we already mentioned in Sect. 3.1 finds a natural explanation here. As an immediate consequence of the entanglement of a person's psychosomatic state, their body keep track of the succession of their emotional experiences, which forms their subjective life. This bodily inscription results from the psychophysical correlations established during the individual's preparation as a biological, psychological and social subject and constitutes their psychosomatic signature. Note that this theoretical justification of the existence of a body memory provided within our systemic approach is corroborated by Damasio's "somatic markers" theory, which is grounded on the idea that the emotions experienced by a person are associated to corresponding changes in the body (Damasio 1994). It is also in accordance with Kandel's theoretical and experimental results

[3] These probabilities are calculated by taking the square module of their components in the psychosomatic state.

according to which a memorization process, which can be that of an experienced emotion, gives rise to associated changes in the neurons of the brain (a functional change for short term memory and an anatomical one for long term memory).

Now taking the time independence of mind-body entanglement into account leads to a second fundamental mode of healing based on *the restoration or the reorganization of the patient's body memory by acting on their body.* As mentioned in Sect. 3.1, this mode of healing has already been explored by some practitioners. Reich suggested that repressed emotions are always accompanied by muscles stiffening, which he calls the "somatic muscular breastplate", and that bodily interventions, such as relaxing muscular tension through massages and breathing exercises could release these emotional sufferings imprinted in the body (Reich 1972). More recently, this mode of healing has been implemented in Fiammetti's somato-emotional approach. In his book, *Le langage émotionel du corps,* this author mentions the psychophysical correspondences originally established by Traditional Chinese Medicine and sometimes refined or reinterpreted according to his own practice, by referring to the cases encountered in his practice (Fiammetti 2004). Fiammetti's method of healing is grounded on the idea that "emotions that are not expressed are impressed in the body" and that decrypting this "body language" thus allows a somato-emotional rebalancing. Similar to this idea, the Feldenkrais method is a bodily approach that uses movement to improve well-being (Feldenkrais 1967). A randomized clinical study of 20 subjects with multiple sclerosis showed an improvement in the rate of depression and anxiety in patients who had used the Feldenkrais method for 8 weeks (Rakel 2007). Indeed, several techniques of "personal development" are grounded on this idea of working on body memory. For example, Mathieu Vénisse's website[4] proposes several techniques to help one's body getting rid of emotional traumas. These techniques include "guided relaxation", which involves the practice of meditation or mental visualization,[5] "global gymnastics", which aims to develop the awareness of postures, gestures and attitudes, and "psychological massages", which aims to make the individual aware of the link between their physical symptoms and emotional trauma. This awareness is obtained by working on the fascias, which are fine membranes that envelop muscles and organs, using a precise technique similar to Fiammetti's somato-emotional therapy. These alternative therapies are thus based on the idea that a patient encapsulates the past emotional conflicts that have not been expressed in some specific regions of their body and that an action on these regions can release them.

The suggested theoretical justification of the existence of a body memory from the mind-body entanglement perspective can moreover be extended to that of a *transgenerational* body memory. A person's body also keeps track of their ancestors' emotional experience because in ontogenesis the development of the three embryonic layers giving rise to organs is correlative to (and shaped by) the archaic emotional conflicts experienced in their embryonic and foetal past through their

[4] https://www.penser-et-agir.fr/

[5] Mental visualization is considered in the next section, which focuses on self-healing techniques.

mother's experience, which in turn relies on all of their family story. This theoretical consequence echoes with astonishing observations of such a transgenerational body memory. For example, by analyzing a subject's genealogical tree showing the ties between their family members and important traumatic events they experienced, Anne Ancelin-Schutzenberger highlighted problematic repetitions from generation to generation (Ancelin-Schutzenberger 1998). She suggested that our body would be a repository of all ailments and wounds of our ancestors and that some of them can appear at specific times and express themselves through our body. An illness, for example, could manifest itself at a specific time in life, in the same way an ancestor experienced it. Mention should also be made of Josephine Hilgard's work (1953) which showed the significant repetition of coincidences between the dates of accident or psychotic decompensation of patients and those of their ancestors.

This second holistic mode of healing thus suggests that working on one's body memory can get rid of the past, even transgenerational, emotional traumas imprinted in it, by first highlighting the relevant areas of the body and then by acting on them. This allows to free the subject from negative or blocked emotional traumas and to "inscribe" in their body new positive and rehabilitating emotional experiences. This method of healing aiming to restore one's body memory requires the knowledge of the patient's psychosomatic signature, that is, the knowledge of the specific psychophysical correlations encoded in their psychosomatic state and their strength. This knowledge reveals which emotional experience the patient associates to specific bodily dysfunctions; it can also reveal some bodily dysfunction of which the patient is unaware. In order to further explore these psychosomatic correlations, we first need to characterize more accurately the somatic and mental entities considered.

A somatic description does not only involve physiological observables, like the cardiovascular observables considered in the meta-analysis of Chap. 4, but all anatomic, immune and endocrine features of the individual. Similarly, as shown in Sect. 4.2, the emotional state of a person involves several interdependent emotions since they are related by complementary relations. Moreover, the dynamics of somatic and mental observables are closely related to each other in order to ensure the synchronization of both biological and psychological rhythms, which is fundamental for the global psychosomatic balance (see Sect. 4.1). These complex correlations between somatic observables, between mental observables and between somatic and mental observables can be described in a synthetic way as correlations between specific *biological programs,* involving the functioning of several organs, tissues and viscera, and specific *emotional tones,* involving several emotions and the way they are connected to each other. What has been prepared during evolution is thus a collection of biological programs related to specific emotional tones generated from the subject's experienced emotions. More precisely, taking the inherent indeterminacy of all the involved somatic and emotional observables into account (property of non-locality of mind-body entanglement), we should say that what has been established is *a set of possible pairs* of the type {emotional tone / corresponding biological program}. Within the quantum-like representation developed in

Chap. 4, a person's psychosomatic state can thus be conveniently represented by a linear combination of product-states encoding the correlations between emotional tones and biological programs[6]:

$$|\Xi >= \Sigma_{a,b}\, c(a,b)|M_a, P_b >,$$

being understood that the "emotional tones" M_a involve several interdependent emotional observables while the "biological programs" P_b involve several interdependent somatic observables. In particular, if some of these emotional tones are traumatic, the corresponding biological programs can be considered as appropriate "responses" of the body to cope with them. This means that within the holistic framework presented here the internal diseases a person develops can be regarded as biological programs realized by their body in association with, or "in response to", some experienced psychological conflicts. Consequently, the knowledge of a person's specific psychosomatic state will enable us to understand the genesis and evolution of the internal diseases they can develop.

Let us now study in more detail the constitution of a person's body memory by referring to their preparation process. This "preparation" process is the evolution process broadly construed: phylogeny, ontogeny and social life. In phylogeny, the establishment of psychophysical correlations is essential to prepare the body to deal with primordial emotions (hunger, thirst, fear of suffocating, etc). The individual specificity of a person's psychosomatic correlations has to be found in their personal history, which begins from ontogenesis and extends into the social life. In this individual preparation process, embryogenesis seems to be the main moment since it associates, via the development of the embryonic leaves, the primordial emotional conflicts to the very manufacturing plan of the tissues forming their organs. As briefly mentioned above, it is in embryogenesis that correlations are established between the foetus' experienced emotions and the development of the three embryonic layers (endoderm, mesoderm, ectoderm) from which specific tissues and organs as well as the tripartite structure of the brain originate. Correlations are then established between primary psychological conflicts and biological programs via the development of the embryo. These correlations have been studied by Hamer, a cancerologist, by using imaging techniques, such as computed tomography or magnetic resonance imaging (Hamer 2005). Hamer has shown the existence of strong correlations between specific emotional conflicts and the dysfunction of specific organs. For example, between an excessive increase in the growth rate of cells in the lungs and a *traumatic, emotional* shock linked to the primordial fear of suffocating. The fear of suffocating would thus be directly correlated with the excessive development of lung tissue, which facilitates a greater oxygen supply to the blood. Highlighting this correlation may help to explain that a recent emotional conflict related to the fear of suffocating (which is also psychologically linked to the fear of dying or being

[6] We use here the writing suggested in Sect. 4.1, grouping the terms of the psychosomatic states so as to emphasize the emotional and somatic features we want to focus on.

abandoned) may lead to overproduction of pulmonary tissues, which could therefore suggest a future avenue to explore in the study of the genesis of lung cancer.

More precisely, Hamer identified the correlations between essential emotional conflicts, echoing archaic conflicts, and the activation of specific "foci" in the brain, which govern the development and the functioning of the organs. He has for example specified that the "old brain", consisting of the brainstem and cerebellum, manages the oldest conflicts such as those related to breathing, feeding, digestion or reproduction. The reason is that the cells of the organs responsible for digestion or respiration, which are essential for the survival of the individual, originate from the endodermal leaflet and depend on the brain stem, which is the oldest part of the brain. Consequently, there will be a strong association between archaic emotional conflicts related to the impossibility of breathing, feeding or swallowing and specific pathologies in the organs governed by the brainstem and derived from the embryonic endoderm. Similarly, Hamer explained that self-devaluation conflicts are expressed through the development of the mesoderm from which originates the limbic system. As a consequence, self-devaluation conflicts are correlated to lesions of the bones, muscles and tendons, which originate from mesoderm, as well as with the activation of brain foci located in the limbic system. The psychological conflicts related to communication and territorial integrity rely on the development of ectoderm, thus correlating with the brain foci located in the cerebral cortex that govern the possible lesions of epidermis, coronary artery and the mucous membranes of several organs, including the larynx, esophagus, nose or mouth.

Note that the previous analysis is corroborated by the efficacy of acupuncture, which was briefly presented in Sect. 3.1, and reflexology. Reflexology consists in the application of pressure to specific points on the feet, hands or ears in order to act on the dysfunction of corresponding organs. Like acupuncture, reflexology is based on a holographic concept of the body that puts in correspondence some areas of the body with its global functioning. In these therapies, the stimulation of specific points on the body allows treating pathologies in other, possibly remote parts of the body, while activating correlatively specific brain areas (Cho et al. 1998). How could these long-distance connections be explained? In spite of interesting assumptions aimed to explain such connections, meticulously reported by Janssen (2006, p. 231–233), it seems that, until now, no convincing biomechanical or biochemical explanation has been provided about an alleged circulation of some energetic fluid, called "qi" in Traditional Chinese Medicine, along the meridians, from the stimulated point to the treated organ. A plausible explanation of the efficacy of these techniques is that, as suggested by Janssen, the connection between brain areas and remote parts of the body, as well as between its parts, could be explained by their common origin: they originate from the same embryonic layer (see the previous paragraph). Clarifications to this idea can be provided within our systemic approach based on the concept of mind-body entanglement: in spite of their spatial separation on the body, the elements of each of these groups of body parts confectioned with the same embryonic stuff have memorized similar emotional experiences.

Consequently, the remote parts of the body originated from the same embryonic layer are connected to the same emotional experiences and, therefore, they are connected to each other through these emotions. For example, in acupuncture, the experienced emotion of fear, the acupuncture points of the kidney meridian and the kidney organ are strongly correlated without any exchange of signal. Consequently, the stimulation of any of the points of the group "fear/organ-kidney/kidney-meridian" can activate the other ones, independently of the distance separating them and of the existence of a hypothetical channel of communication between them. This means that the previous implementation of the concept of mind-body entanglement in embryogenesis could justify the holographic model of the body on which acupuncture and reflexology are grounded.

Turning now to our main purpose, it can be asserted that the knowledge of these specific psychophysical correspondences can explain those between emotional conflicts and bodily dysfunctions. This knowledge can thus explain the holistic origin of a person's internal disease and lead us to the development of appropriate ways of healing on both psychological and somatic grounds. For example, in the previous example of lung cancer, a psychological follow-up focused on the problem of abandonment could be really essential, in complement of the usual drug and radiation treatments which can be necessary to stop its proliferation in the short term. Thus, in contrast with the current therapies proposed, whose action generally consists in exerting a local action on the "diseased" tissues, our holistic approach to illness does not only focus on the dysfunction of a particular organ, such as the lung in the previous example, but also takes the patient's psychosomatic organization into account. It regards the local damaged tissues or a specific organ's dysfunctions only as *symptoms* or *signs* of the global, psychosomatic imbalance. This holistic approach to illness can be studied along the scientific method: one can use medical imagery and physiological analysis to control the evolution of the somatic symptoms of a sick person and follow, at the same time, the evolution of their emotional state through psychotherapy or psychoanalytic analysis, or, more directly, by using questionnaires (like in Sect. 4.4). It extends the line of thought upheld by Hippocrates, by psychosomaticians such as Alexander and Groddeck (see Sect. 3.1) and, more recently, by researchers in psycho-neuro-endocrino-immunology (Goodkin and Visser 2000); it also conforms to the spirit of Traditional Chinese Medicine. It leads to the development of an alternative, holistic medicine that focuses not on a disease's somatic symptoms but on the patient's psychosomatic organisation. Along this systemic approach, healing a patient first consists in looking for the specific psychophysical correlations that characterize their psychosomatic organization and then acting on the relevant mental and somatic features, directly or through the restoration of their body memory, in order to rebalance their psychosomatic state. Other illustrations of this holistic approach to illness will be provided in the following chapter.

5.3 Self-Healing Processes: Placebo "Effect", Biofeedback and Mental Imagery, Meditation

The possibility of healing one's body by reaching specific mental states has been briefly mentioned in the previous section, as a direct mode of healing. If self-healing methods have been known and practiced for a long time, they remain totally unexplained within the classical mode of thinking, based on the existence of a hypothetical causal interaction between mind and body. This is particularly the case regarding the placebo and biofeedback phenomena, as well as the practice of meditation. As shown hereafter, the possibility and the efficacy of such *self-healing* techniques can be explained from mind-body entanglement, which makes the psychosomatic unity of the individual possible.

5.3.1 The Placebo "Effect"

Double-blind trials have shown that pharmacologically inert substances can produce the same physiological effect as genuine pharmacologically active substances (such as a pain-killer or an antibiotic). This therapeutic effect of a pharmacologically inert substance such as a sugar tablet or a saline solution is called "placebo effect". According to the pathologies, the placebo effect can heal or, at least, ease the symptoms of a disease, for 10–90% of the cases. For example, it has proven to be very successful for relieving pain, eradicating a cough, healing a gastric or duodenal ulcer, fighting against hypertension and even for treating Parkinson disease (Janssen 2006; Fuente-Fernandez et al. 2006). The somatic effects of a placebo have been well established by observing, for example, changes in the body temperature, blood pressure or heart rate, a decrease of the rate of gastric acidity or cholesterol level. For the treatment of pain, it has been shown that the placebo phenomenon activates the same regions of the brain that are stimulated by opioid analgesics and triggers the release of endorphin by the brain (Petrovic et al. 2002). The placebo "effect" could thus a priori be used as a therapeutic resource and not only be considered as a nuisance or background noise whose role is only to test the efficacy of drugs -as is generally the case in contemporary pharmacology[7] and medicine.

However, if the placebo's somatic efficiency is real, it nevertheless relies on pure psychological factors, essentially the patient's faith in the physician and in the prescribed treatment, and, secondary, on its form (an injection being generally more successful than a pill), its size or its mark (Janssen 2006; Peters 2001). This means that purely psychological factors, which are of the order of suggestion or belief,[8] can

[7] Clearly, this attitude regarding the use of the placebo "effect" has an important financial motivation: if placebos can be used to heal, why should we take (and pay for) drugs?

[8] The specific role of belief in healing processes, or rather of the human capacity of believing, will be analysed in more detail in Sect. 5.5, which addresses distant healing phenomena.

produce significant (and measurable) changes of physiological factors and then significant improvement of the global state of a patient. Looking for an explanation of the placebo effect leads to the long-standing question of the nature of the relationship between mental and physiological processes (see Chap. 1). This question remains unanswered within the causal mode of thinking, since the latter cannot explain how the patient's sate of optimism and confidence would be causally responsible for triggering the biochemical and neural effects that mimic that of a real treatment. The idea according to which the mental could merely "cause" a change in physiological variables relies on a concept of mental causation which, as explained many times in this book from different approaches, is not appropriate to deal with the mind-body connection (see Chapts. 1–3). However, the placebo phenomenon can find a natural explanation within the systemic, quantum-like framework which interprets the mind-body connection as a relation of generalized entanglement, that is, without appealing to any concept of psychophysical causation.

Representing the psychophysical correlations in terms of generalized entanglement allows the association in the same "product-state" of restorative somatic processes *and* positive emotions, correlations established by evolution. This can naturally explain that stimulating a mental state of confidence in the doctor's words and in the prescribed treatment can correlatively activate the repairing processes of the body. In such a setting, the placebo effect formally consists of a modification of the amplitudes of the different components of the psychosomatic state. The probabilities of occurrence of "positive" product-states governing positive emotions and restorative somatic processes, which were almost null initially, increase while those of "negative" product states, associated with negative emotions and pathological processes, decrease.

In order to illustrate this approach to the placebo phenomenon, let us consider the very successful use of a placebo to release pain. The basic somatic states of space S will be defined by the values of observables that measure the activation of the regions of the brain involved in pain, the sympathetic and the parasympathetic nervous system's activity and the endorphin secretion rate of the brain. The basic mental states of M will be defined by the values of the mental observables that evaluate the patient's trust in their physician and in the prescribed medical treatment. Adopting once again, for sake of simplicity, the synthetic representation of psychosomatic states introduced in Sect. 4.1, we will consider that these somatic and mental properties can be encoded in only two state vectors of the somatic and the mental spaces respectively. The psychosomatic state of a subject will thus be written as:

$$|\Xi\rangle = a_+ |\phi^+ \psi^+\rangle + a_- |\phi^- \psi^-\rangle,$$

with $|a_+|^2 + |a_-|^2 = 1$.

In the latter expression $|\phi^+\rangle$ and $|\phi^-\rangle$ respectively refer to somatic states for which the pain is either intense ($|\phi^+\rangle$) or weak ($|\phi^-\rangle$). As relevant physiological variables one can use the patterns of brain activity, such as those of the somatosensory cortex or the insular cortex, or, regarding the whole body, the level of tension of some muscles or the excitation rate of specific nerves. In the same way, $|\psi^+>$ and $|\psi^->$ denote, respectively, the correlative "positive" and "negative" mental dispositions regarding the physician and the prescribed treatment, as well as any other psychological factor involved in this situation that can be evaluated, for example by reporting its level on a scale. The entangled psychosomatic state $|\Xi>$ written above is defined as a linear combination, weighted by the complex coefficients a_+ and a_-, of the product-states $|\phi^+\psi^+>$ and $|\phi^- \psi^->$.

The initial state $|\Xi>_{in}$ in which the patient feels strong pain can be described by a very weak value of the probability associated with amplitude a_+: $|a_+|^2 \cong 0$, which could signify that their brain does not produce a sufficient amount of endorphins capable of relieving this pain. Now, an optimistic suggestion from a trusted physician can activate the positive feelings that define the mental state $|\psi^+>$. Due to the entangled form of the psychosomatic state $|\Xi>$, $|\psi^+>$ is strongly correlated to $|\phi^+>$ in the product $|\phi^+\psi^+>$, which means that the positive somatic correlates of $|\psi^+>$ that define the somatic state $|\phi^+>$ are also activated *correlatively*. The probability $|a_+|^2$ will then increase while, by the normalization condition, the « negative » probability $|a_-|^2$ will decrease and, in the ideal case, tend to zero: $|a_{-fin}|^2 \cong 0$. The placebo "effect" is thus formally expressed by a change in the probabilities of occurrence of the product-states involved in the patient's global psychosomatic state.

The placebo phenomenon can thus be regarded as the paradigmatic expression of the mind-body entanglement which grounds the psychophysical unity of the individual. It is clearly a *self-healing* resource since by reaching an appropriate mental state of optimism and confidence the patient is *by himself* capable of triggering somatic restorative processes. Its efficacy depends of course on personal factors: i.e. (1) the patient's ability to reach a state of optimism and confidence, which relies on the strength of their belief in their physician and in the prescribed treatment; and (2) the intensity of the patient's specific correlations between their "positive" somatic and mental states,[9] which make the triggering of healing processes possible.

5.3.2 *Biofeedback and* Mental Imagery

As mentioned in Chap. 3, biofeedback is a mind-body technique where the subject visualizes the real-time plot of a generally non conscious-accessible physiological variable on a screen so that they can modify it by their mental activity (Chiarioni and Whitehead 2008; Kranitz and Lehrer 2004). More precisely, in biofeedback electrodes are placed (on the skin) on places of the body allowing the measurement of the chosen variables (for example, those measuring a muscle's contraction, heart rate or brain's activity) and the electric signal is sent to a monitor that displays the

---

[9] As explained in Sect. 4.1, these somatic and mental states are respectively defined from the values (eigenvalues) of the mental and somatic observables involved in the required healing processes. In the previous example of pain, a relevant somatic observable measures the brain production rate of endorphins while a relevant emotional observable measures the intensity of happiness.

changes of these variables. In order to give comprehensible information to the patient so that they can rapidly understand these physiological signals and control them, the electrical variations measuring those of the chosen physiological variables are converted into audible signals (for example, by changes in pitch) or in visual signals (for example, by changes in brightness). The biofeedback technique has been successfully used for releasing stress (its initial goal, indeed), to treat migraine headaches, heart disease, sleep disorders, muscle dysfunction, respiratory problems and many other disorders (Mullally et al. 2009; Hammer et al. 2011; Moravec et al. 2011; Kubik and Biedroń 2013). Note that biofeedback relies on the sole visual sense but it is believed that, more generally, a subject can use other senses, like touch or smell, in order to obtain the expected bodily changes (Barber 1978). The general ingredient of such a "mind-to-body effect" is mental imagery, defined as "any thought representing a sensory quality" (Horowitz 1983) and playing an essential role in all human mental tasks. On biological grounds, it has been shown that mental imagery could function like afferent sensory perception, for example by inducing an orientation effect after imagining oriented lines (Mohr 2011) or inducing a motion effect on a subsequent perceptual stimulus after imagining a moving stimulus (Winawer 2010). Biofeedback, and more generally mental imagery, are thus techniques helping the subject to reach the mental strategies or states correlated with the expected bodily changes. Like for the placebo phenomenon, they can be regarded as self-healing techniques since it is the subject himself who modulates some chosen physiological functions by its own mental activity. As mentioned above, this self-healing technique can prevent or treat a range of health problems in addition to medication, or even without it.

How can this mental "control" of physiological variables be possible? We are again led to the question of the nature of the mind-body connection and, particularly, that of mental causation. Like for the placebo phenomenon, an explanation of biofeedback and mental imagery can be provided within the systemic, quantum-like approach presented here. This "effect" of the mind on the body can be explained by referring to the psychophysical correlations encoded in the patient's psychosomatic state, without evoking a mysterious causal interaction. These correlations have been established during the evolution process broadly understood.[10] The expected somatic changes are obtained by enhancing the mental states that are positively correlated with them. For example, lowering the heart rate, blood pressure or brain's generation of theta waves are due to (and are not mechanically "caused" by) the subject's mental capability to reach a state of quietude, which lowers their psychological stress. The transition to such a mental state is correlated with the previously mentioned changes in physiological variables in their psychosomatic state. The mathematical representation of biofeedback and, more generally, of mental imagery is thus quite similar to that of the placebo phenomenon analyzed above: the probabilities of the targeted physiological states are increased correlatively to those of the

[10] Recall that if these correlations have been established during the subject's history, they are not fully determined by this history due to the fundamental indeterminacy of mental and somatic observables (see Sect. 4.2).

mental states reached by the subject because these two somatic and mental states are correlated in their entangled psychosomatic state.

5.3.3 Meditation

If biofeedback uses monitoring devices to help the subject reach appropriate mental strategies or mental states in relation with the expected bodily changes, such a use is indeed not essential. Experimental studies have shown that people trained in meditation can control their physiological parameters without the help of monitoring devices. For a very long time, different forms of meditation have been developed to teach the subject to navigate in the inner space of conscious states. Meditation, whose origin lies in the Indus Valley 4000 years ago, was originally conceived of as a mental and spiritual practice aiming to reach a state of inner peace and spiritual awakening (Janssen 2006, Chap. 4). It has also been regarded as a particular mode of access to the knowledge of reality (Rommeluère 2015). For our purpose, we will focus on the self-healing effects of meditation, which have been known for a long time and have been studied using scientific methodology. Meditation practices can have two forms depending on whether or not the meditator concentrates on a specific idea, object or sentence. Regarding the first form, let us mention Transcendental Meditation, developed by Maharishi Mahesh Yogi (1963) from ancestral spiritual Indian practices. This paradigmatic form of meditation asks the meditator to concentrate on a specific sentence, a "mantra", repeated mentally while sitting in a comfortable position. Its healing "effect" has been studied by Herbert Benson (1975) under the name of "relaxation response". Benson's relaxation response consists in measurable indicators of stress reduction, such as the lowering of blood pressure and oxygen consumption, of heart and respiratory rate, and of the central nervous system activity -all these indicators being indeed involved in the first phase of sleep (Benson 1975). Experimental studies have shown that the practice of Transcendental Meditation can significantly reduce high blood pressure or cholesterol level (Cooper and Aygen 1979), as well as alleviating anxiety (Benson 1975) and depression (Ernst et al. 1998) in the mental domain,. These experimental studies thus show that Transcendental Meditation can be very useful for preventing or healing various somatic dysfunctions or mental disorders.

The second form of meditation, called Mindfulness, has been taught for over two millennia. This type of meditation does not suggest to concentrate on a particular item but involves "focusing your attention in a particular, deliberate way, in the present moment and without value judgment" (Kabat-Zinn 1994). In 1979, Jon Kabat-Zinn, a doctor in molecular biology and practitioner of yoga and meditation, suggested to adapt the ancestral practice of meditation to the hospital-environment, where stress is very present. By focusing on the conscious experience of every moment, the practice of Mindfulness allows a significant reduction of negative emotions, while maintaining positive ones. Similar to that of Transcendental Meditation, its effectiveness has been demonstrated in many healing processes, as for example for treating pain in weakening the subject's response (Petter et al. 2014), for reducing the breathing rate or lowering blood pressure and then reducing the risk of heart attack (Schneider et al. 2005). In the mental domain, Mindfulness has proven very successful for releasing psychological stress, depression and anxiety (Hofmann and Gómez 2017). Like for the other forms of meditation, this efficacy relies on the brain's plasticity, that is, its ability to be physically modifiable, thereby reinforcing the cerebral regions and pathways that correlate with the generation of positive emotions, while weakening others (Laureys 2019). For example, brain imagery has shown that regularly practicing meditation for 2 months significantly increases the activity in the anterior part of the brain's left hemisphere, which is associated with experiencing positive emotions (Davidson et al. 2003). An experimental study in which the Buddhist monk Matthieu Ricard participated has shown an increase of brain gamma waves and of the left prefrontal brain activity with time spent meditating (Lutz et al. 2004).

So, even without necessarily appealing to external control devices, the practice of meditation aims to generate a state of consciousness capable of calming the mind and relaxing the body. As was the case for the placebo effect, for biofeedback and mental imagery, the practice of meditation can thus be regarded as a *self-healing* technique. Its efficacy, which is reinforced by the practice, depends on the meditator's capability to self-regulate its attention in order to reach appropriate mental states, and, through mind-body entanglement, to trigger the correlated healing bodily processes. As was the case for the self-healing techniques previously mentioned, this efficacy relies on the subject's own psychosomatic structure, which results from their long phylogenetic, ontogenetic and social history.

Let us end this section by mentioning self-healing phenomena realized through prayer. A prayer assigned to its own well-being can be regarded as a form of meditation where the meditator concentrates on a specific idea (healing oneself). Along this line of thought, many self-healing phenomena can be mentioned, generally related to religion and qualified by religious and medical authorities as "miraculous healings." Such miraculous healings have been reported as occurring after praying to religious authority, like Virgin Marie (Lourdes' miracles), Jesus or, closer to us, the Pope. For example, Sister Marie Simon-Pierre was miraculously cured of Parkinson's disease on the night of June 2, 2005, following her insistent prayers to John Paul II who died on April 2, that is, *two months earlier*. This means that it is much more plausible to think that this miraculous healing is due to Sister Marie Simon-Pierre's *own prayer* rather than to John Paul II's healing power because he was not alive at the time. We can also mention the case of Floribeth Mora Diaz who was diagnosed with serious brain damage (rupture of aneurysm in the right cerebral artery with hemorrhage) and told she only had a few hours to live. She reported that she was looking for comfort in God and prayed through the intercession of John-Paul II. One day, when she was alone in her room, she heard an inner voice inviting her to get up. She reached a state of great peace and self-confidence and was no longer afraid of dying. This emotional state led to the miraculous healing of her supposedly incurable stroke.

Like the placebo effect, biofeedback and meditation, these cases of miraculous healings, as many others reported for example by Lunel (2004), could thus be interpreted as self-healing phenomena. Reaching "positive" emotional states, such as those of optimism, empathy and confidence (in the placebo phenomenon), concentrating on mental images and strategies (in biofeedback), reaching a state of peace and spiritual awakening (in meditation) or devotion to a religious authority (by praying), can trigger body healing processes.

5.4 Psychiatric Domain

Like "physical" diseases, psychiatric disorders can better be understood within the systemic, quantum-like framework presented here. New ways of healing that complement drug medicine, which often has partial efficiency and secondary effects (Zahbib et al. 2012, Sect. 4), can be proposed. For example, bipolar disorder is characterized by a high probability of occurrence of only two states (or even only one in its unipolar version), a manic or hypomanic state and a depressive state, while the other product-states composing the global psychosomatic state of a person have almost a null probability of occurrence. Hence the elaboration of a model of bipolar disorder where the psychosomatic state of the patient can be approximated by a two-states entangled system. As will be explained below, such a model provides fruitful responses to some important unresolved questions relative to this psychiatric disorder's genesis, organization and evolution (Bourgeois 2014). Another example of mental disorder that can greatly benefit from this holistic approach is schizophrenia. Schizophrenia has been characterized by Bleuler (Patouillard 2011) as a disorder of the associative function resulting in a dissociation (Spaltung) of psyche functions. This idea of psychic function dissociation can be represented according to the holistic approach developed here. It is expressed by the "cut" between some of the possible elementary product-states composing the psychosomatic state of the individual, that is, by *the quasi-impossibility of transition between them.* This means that their probabilities of transition are very small, as if these elementary states were mutually exclusive. For both disorders, instead of focusing on their symptoms and treating them by psychotic medications, the healing process would consist in rebalancing the psychosomatic state. This can be done by re-initializing its components, which are probability amplitudes, and reinforcing the transition probabilities between them. Such a way of healing will be illustrated below for bipolar disorder.

Before presenting below an explanatory and predictive model of bipolar disorder within the systemic, quantum-like approach explored here, we will first mention some general consequences of this approach with regard to long-standing fundamental questions about mental disorders. These consequences deal with the question of the alleged somatic causation of mental disorder, with the distinction between normality and pathology as well as with the distinction of mental disorders on which is based their classification in the *Diagnostic and Statistical Manual of Mental Disorders* (DSM).

5.4.1 No Somatic Causation of Mental Disorders

An important consequence of the concept of mind-body entanglement is that the notion of a somatic "cause" of a mental disorder has no meaning. The somatic and the psychic aspects of the psychosomatic unity of the individual are their *two complementary descriptions* that are not causally connected despite their strong correlations. An illustration of this idea has been provided by Fig. 2.2 (Chap. 2) where the movement of a single bird observed under two different angles gives rise to two strongly correlated aspects that are yet not causally connected because no signal is sent from one of these aspects to the other. Turning now to the psychiatric field, this means that a brain dysfunction cannot be considered to be causally responsible for a mental disorder even though correlations have been observed between this dysfunction and some symptoms of this mental disorder. For example, contrary to what has been argued by the American Psychiatric Association (2010), schizophrenia cannot be considered as a "brain disease" due to a chemical imbalance, such as an excess of dopamine (Healy 2008). Similarly, Johnstone, Frith, Crow and Owens's observation (1988) that schizophrenia is significantly associated with an increase in the volume of the lateral ventricles of the brain (which is indeed observed in many other cases (Read 2004)) does not prove at all the existence of a causal relation between brain dysfunction and schizophrenia. And the heredity of schizophrenia, which has been rightly noted (NIMH 2010), does not show that schizophrenia has a "genetic cause", even by cautiously qualifying this disease as "a complex genetic disorder" involving many genes and sensitive to environmental factors (Gejman et al. 2010). Indeed, as is well known, heredity does not necessarily imply Mendelian genetic transmission and the observed heredity of schizophrenia could a priori have environmental and familial factors rather than genetics. On the contrary, and as explained throughout this book, interpreting the mind-body connection in terms of causation seems totally inappropriate: *no consistent concept of psychophysical causation can be built*, hence the claim that there is *no somatic causality of mental disorders.* As explained, the property of a-causality of the mind-body connection is captured by the concept of mind-body entanglement.

The assertion according to which there is no somatic causality of mental disorders has been experimentally confirmed by the negative results of statistical studies conducted to assess the degree of correlation between schizophrenia and brain dysfunction, and between schizophrenia and genetic mutations. These statistical studies, presented in Williams' book (2012, Chap. 2), show that no significant positive correlation has been observed, neither between schizophrenia and brain dysfunction, nor between schizophrenia and genetic mutations. Insofar as the existence of a causal relationship implies the existence of positive correlations between the supposed "cause" and the supposed "effect", the absence of such positive correlations implies the absence of any causal relationship. In addition, Williams has stressed that some statistical studies supposed to show the existence of a "genetic causality" of schizophrenia are tainted with several methodological biases, in particular the neglect of the role of the environment in which the couples of twins (monozygotes vs dizygotes) were born and raised, and for which the frequencies of coincidence of schizophrenic involvement are compared (Williams 2012 p. 26–30).

5.4.2 The Continuum of Psychosomatic States: Continuity Between "Normality" and "Pathology" and Between Mental Disorders

A consequence of the representation of psychosomatic states by linear combinations of psychophysical product-states whose (real and imaginary parts of) components vary *continuously* in the set of complex numbers is that the space of psychosomatic states is a *continuum*. More precisely, these states form a continuous partial order since the components of two different states cannot always be compared by the usual order relation between real numbers. The latter, which ordered two real numbers, can only be used to compare two states whose components are all identical with the exception of one of them, and for which only the real parts (or only the imaginary parts) are different. Consequently, if one adopts this representation no clear border can be drawn between "normal" states and "pathological" states since they do not respectively refer to discrete entities that could be clearly compared.

An ideal notion of "normality" could nevertheless be roughly defined by referring to the state of psychosomatic balance. Such a "normality" could for example be characterized by the class of states for which the square modules of its components (which are their probability of occurrence) remain closer to each other and the transition probabilities between them keep a minimal value. These conditions respectively mean that the contribution of all basic psychophysical product-state are balanced and that they are all easily accessible by the subject. This notion of normality seems to correctly interpret the idea of psychosomatic balance already explored by Greek physicians and by traditional Oriental philosophies and medicines, and psychosomaticians (see Sect. 3.1). While for the class of more "pathological" states there would be greater differences between the probabilities of occurrence of the basic product-states and/or the transitions between product-states would be less likely, if not impossible. The example of schizophrenia above and the model of bipolar disorder to be developed below seem to illustrate these definitions.

Note that, correlatively, a sharp distinction between *internal*[11] physical diseases cannot be maintained either, since the somatic aspects of psychosomatic states, which are projections (in the mathematical sense) of psychosomatic states onto the subspace of somatic states, also form a continuous partial order. This consequence agrees with Claude Bernard's idea of a continuity between the normal and the path-

[11] We stress the word "internal" to say that, as argued by Canguilhem (1966), the idea of continuity between the normal and the pathological might not apply to infectious diseases where an exogenous microbial agent penetrates the body and could then induce a qualitative change of state.

ological (Bernard 1876). Claude Bernard considered a disease not as a new state, qualitatively different from the normal one, but only as a quantitative variation of a normal process or a simple deviation from a norm.[12]

Similarly, the continuous partial order formed by the set of psychosomatic states makes it very difficult to distinguish mental disorders form each other and to classify them (DSM). For example, where does the class of states listed as "schizophrenic", "paranoid" or "bipolar" begin and end in this continuous partial order? On which criteria would these distinctions be based? As was the case for the suggested definition of "normality", one can conventionally define discrete regions of the continuous multi-dimensional tree structure of psychosomatic states and assign standard denominations to them. However, nothing ensures that these discrete regions of the psychosomatic partial order would refer to distinct groups of symptoms. This difficulty in distinguishing between typical groups of symptoms has been evidenced by the results of a statistical study reported by the British Psychological Society and quoted by William (2012, pp. 36–37). This research investigates "whether people's psychotic experiences do cluster together in the way predicted by the diagnostic approach." Actually, no sharp distinction can be established between groups of symptoms developed by patients diagnosed with different mental disorders.

5.4.3 An Explanatory and Predictive Model of Bipolar Disorder

Let us now present an explanatory and predictive model of bipolar disorder constructed along the systemic, quantum-like approach presented here. Of course, as explained above in this section, it is often difficult to assign a patient's symptoms to a specific, well-defined psychiatric disorder, namely that of bipolar disorder, since these symptoms are often experienced by patients suffering from other listed disorders, such as depression. It is then rather difficult to diagnose "bipolar disorder" with certainty. However, it seems that in first approximation such a diagnosis can be provided on the basis of *the alternation* of depressive and manic episodes whose symptoms are well documented. Historically, if the existence of such "extreme moods" of a unique continuous mood spectrum had already been mentioned by ancient Greek physicians, the clinical observation of their alternation has first been reported by the alienist Jules Baillarger (1854), who termed this disorder *"Insanity of Double Form"*, and by Jean-Pierre Falret, who gave it the name of *"Circular Insanity"*. It is only at the end of the nineteenth century that Emil Kraepelin (1921) introduced the name of *"Manic-Depressive Insanity" to designate this disorder.* We

[12] The existence of such a "norm" of functioning of the human body has been later captured by Canon's concept of homeostasis to designate the tendency of the organism to restore the balance of its internal environment.

will concentrate on its main features, and first of all on the alternation between manic and depressive episodes.

Bipolar disorder is characterised by the unusual shifts in mood much more severe than those that a "normal" individual can experience. In spite of the great number of studies on bipolar disorder,[13] it seems that some important questions concerning its genesis, its organization and its evolution, do not have clear and precise answers. For example, how can one explain the overinvestment of only two states (mania or hypomania and depression) and their possible superposition,[14] or understand the existence of only a single state (depression) in the unipolar version of this disorder, even though the mood of a "normal" individual can change continuously within a wide range of states according to their experience, personality and environment? How can be explained the existence of so severe fluctuations in mood between depression and mania? Is it possible to predict (and then to prevent) the moment in which one state shifts to the other? Can we understand what triggers this shift? Modelling bipolar disorder according to our systemic approach makes it, at least partially, possible to answer to these questions. In addition, it can provide quantitative predictions and propose new forms of therapies.

Four types of bipolar disorder have been identified by the American Psychiatric Association (DSM-IV-TR, 2000) according to the severity of their phases, the rapidity of their shifting, and the existence of mixed states: type I bipolar disorder involves periods of severe mood shifting from mania to depression; type II bipolar disorder involves milder forms of mania (called hypomania) that alternate with periods of depression; cyclothymic disorder involves periods of hypomania and brief periods of depression; and "bipolar disorder not otherwise specified" (BD-NOS) refers to the fact that the patient suffers from some symptoms of the bipolar spectrum without clearly belonging to one of the three previous types. Moreover, the specific form of this disorder a patient contracts and its evolution closely depend on their own psychosomatic structure, which can be roughly characterised by their "personality" and their ability to cope with reality. Individual trajectories of bipolar patients are of course all different when they are accurately observed. However, we will present here a model of this mental disorder that accounts for *its basic features*, which historically distinguished it from other mental disorders (think about the name of "circular insanity" attributed by Falret), namely, the alternation of manic and depressive episodes, the severity of this mood shifting and, in some cases the possibility of rapid-cycling episodes. This basic model can of course be refined to be adapted to the specific forms of this disorder contracted by patients.

The proposed model considers that the psychosomatic state of a "normal" person is balanced, which means that none of the probabilities of occurrence of all possible product-states are zero and that the transitions between these product-states remain possible. There is no overweighting or fixation on one of these states. All product-states, formed by elementary mental states (like introverted, extroverted, anxious,

[13] See, for example, Bourgeois (2014), Mirabel-Sarron and Leygnac-Solignac (2009).

[14] Manic and depressive symptoms can be experienced at the same time in this "mixed state".

depressed, euphoric, ...) *and* corresponding somatic states (defined by the values of the relevant somatic observables), can thus be invested with comparable probabilities and with a relatively low energy. Accordingly, in this model of bipolar disorder, the psychosomatic state will be written as the linear combination of only two states that are over-invested: a manic state, ranging from hypomanic to acute, represented by the vector $|\Phi_M>$ of the space $\mathbf{H} = \mathbf{S} \otimes \mathbf{M}$ of psychosomatic states; and a depressive state, ranging up to melancholy, represented by the vector $|\Phi_D>$ of **H.**[15] The manic state $|\Phi_M>$ is characterized by psychomotor over-excitement, psychic hyperactivity, flight of ideas, euphoric exaltation and insomnia. It generally gives rise to delusions and hallucinations. In contrast, the depressive state $|\Phi_D>$ is characterized by deeply sad and pessimistic mood, inhibition of the psychic and motor functions, anorexia, lack of energy and hypersomnia. These two states can be respectively defined by the values of the relevant mental and somatic observables.[16] We will not deal here with the possible hereditary and social factors that could explain the over-investment of these two states (or even of one of them). We will only write the typical form of the entangled psychosomatic state of a patient suffering from bipolar disorder and focus on its temporal evolution from which predictions can be made. This state can be written as:

$$|\Xi>_{MD} = c_M|\Phi_M> + c_D|\Phi_D> + |\varepsilon>,$$

where $|\Phi_M>$ and $|\Phi_D>$ are psychosomatic states of H encoding, through the values of the relevant somatic and mental observables, the characteristic features of depression and mania described above, while $|\varepsilon>$, which will be ignored in the following, groups all the terms of $|\Xi>_{MD}$ whose modules of complex components are small compared to those of $|\Phi_M>$ and $|\Phi_D>$ and then do not contribute significantly to the psychosomatic dynamics of the patient. The states $|\Phi_M>$ and $|\Phi_D>$ have the following form:

$$|\Phi_M> = |s_M, m_M>,$$

where s_M and m_M denote the somatic and mental characteristics, respectively, mentioned above for the manic state, and.

$$|\Phi_D> = |s_D, m_D>,$$

[15] Let us stress the fact that the states $|\Phi_M>$ and $|\Phi_D>$ are *psychosomatic* states, that is, weighted sums of mental *and* somatic product-states. These states are defined according to the method presented in Sect. 4.1, i.e. by grouping the terms so as to focus on the specific features of these states described below.

[16] Concerning the somatic correlates of the different phases of bipolar disorder, one can mention the hormonal, immunologic, and autonomic nervous system dysregulations observed by Dehaene (Dehaene 1997, Chap. 9), which can give rise to serious somatic diseases (Zahbib et al. 2012).

where s_D and m_D denote the somatic and mental characteristics, respectively, mentioned above for the depressive state.

The transformation of a well-balanced psychosomatic state to a truncated state such as $|\Xi>_{MD}$ means that several components of the psychosomatic state became negligible, namely those of product-states whose relevant observables are not involved in the description of the states $|\Phi_M>$ and $|\Phi_D>$. As explained in Sect. 4.1, the smallness of these components results from the desynchronization of the biological and psychological rhythms with each other and with the natural, night/day rhythm, which can thus give rise to beat or interference phenomena. This theoretical explanation of the smallness of some of the components of the psychosomatic state is corroborated by clinical evidence that bipolar patients actually have their internal rhythm sleep/wake offset from the natural rhythm night/day (Takaesu 2018).

First note that in the case the probabilities of occurrence $|c_M|^2$ and $|c_D|^2$ of the two states $|\Phi_M>$ and $|\Phi_D>$ are close to one other, patients can experience both depressive and manic symptoms at the same time, which consequently explains the existence of the "mixed states" mentioned above. As to the dynamics of the psychosomatic states, Sect. 4.1 has shown that it can be described by the introduction of two operators: a "rhythmic" unitary operator, called **R**, that encodes the intertwining of the periodic variations of biological and psychological variables, which are observed and studied by chronobiology and chronopsychology (Reinberg and Smolensky 1983; Finger 1982; Lestienne 1973)[17]; and an interaction operator that encodes the possible physical, biological, social and cultural interactions of the person with its environment. The temporal evolution of the psychosomatic state of a patient suffering from this psychiatric disorder will now be computed by modeling this state as *a two-level state governed by a harmonic law of evolution (imposed by the circadian rhythm) and submitted to external disturbances.* The rhythmic operator encodes the synchronization of all internal rhythms with the circadian rhythm and the disturbances are due to the patient's biological and social interactions of the patient with their environment as well as to other harmonic operators whose action is much weaker than that imposed by the circadian rhythm. Using the well-known techniques of standard quantum theory (Cohen Tanoudji et al. 1973) allows us to establish the following results (Uzan 2014b)[18]:

1. If the perturbation is independent of time, which means that it can be supposed to be constant, the system will oscillate spontaneously between the two states $|\Phi_M>$ and $|\Phi_D>$. Let us call this first "natural" period of oscillation t_0.
2. In the case where the perturbation is periodic, with a period Δt much smaller than t_0, let us assume it to be harmonic in first approximation and without loss of

[17] For example, the rhythmic operator encodes the rhythmic change of the hormones' concentration in blood.

[18] Let us stress the fact that the following results can be established on the essential condition that the psychosomatic state can be written as a non-factorable *superposition* of the states $|\Phi_M>$ and $|\Phi_D>$, which is the basic assumption of our quantum-like approach.

generality.[19] Its effect is to disrupt internal rhythms, such as internal circadian rhythms (sleep / wakefulness, melatonin secretion, cortisol production, ...) -a disruption which has indeed been observed for bipolar patients (Lyall et al. 2018). This periodic perturbation gives rise to forced oscillations since it imposes a rhythm of oscillation to the components of the overall psychosomatic state. In particular, if the imposed frequency is close to the frequency of the "spontaneous" oscillations ($1/t_0$) calculated in the previous case (for a time independent perturbation), there will be a phenomenon of resonance, which means that the contribution of the states $|\Phi_D>$ and $|\Phi_M>$ in the psychosomatic state of the patient *will become still more important*. This phenomenon of resonance could therefore explain the episodes of acute crisis that has been observed (Bourgeois 2014).

3. In the more realistic case where the perturbation is written as the sum of a constant term, giving rise to a "natural" periodicity t_0, and a periodic term of period Δt much smaller than t_0 it is possible to establish the following mathematical relation linking these two periods and the average period T of alternation between $|\Phi_M>$ and $|\Phi_D>$ (see Appendix 5):

$$T = t_0^2 / 4\pi^2 \Delta t$$

This relation gives rise to testable predictions since we can, for example, calculate the average duration T characteristic of the alternation between the states $|\Phi_M>$ and $|\Phi_D>$ from estimates of the two other parameters, t_0 and Δt, and we can thus anticipate the moment of transition in the new phase of the disease. In order to assess the validity of this formula, we have to compare the calculated average period T with the periodicity really observed. According to the work of Olwen (2007), it can generally be considered that the "spontaneous" periodicity of symptoms for patients suffering from bipolar disorder is approximately equal to $t_0 = 1$ year, which characterizes the circannual rhythm. This circannual rhythm thus results from constant perturbations of the natural circadian rhythm. Moreover, let us assume that a patient is sensitive to the 29-days lunar cycle (whose correlation with the female's menstrual cycle and mood is well known), this cycle being treated as a supplementary harmonic perturbation of period $\Delta t = 29$ days. We then calculate by the previous formula that the period T of alternation between states $|\Phi_M>$ and $|\Phi_D>$ is 115 days. Consequently, under the effect of this second harmonic perturbation, with periodicity 29 days, the rhythm of alternation more than tripled and approached rapid cycling observed during episodes of crisis for this disease and reported, for example, in DSM IV-TR 2000 (pp 490–491).

Let us now consider the possible therapies of bipolar disorder. Treatments for this mental disorder essentially consist in prescribing mood stabilizers, neuroleptics and antidepressants that affect the production and transmission of certain

[19] Since one can always decompose a periodic function in a weighted sum of harmonic functions by the Fourier theorem.

neurotransmitters, as well as the neural and synaptic plasticity (Schloesser et al. 2012). However, as know by psychiatrists, these drug treatments can be partially effective in acute seizures and insufficient for long-term prevention. Moreover they generally have significant adverse effects, such as digestive disorders, tremors or thyroid dysfunction (Bougeois et al. 2014). In contrast, the systemic model of bipolar disorder presented here suggests alternative or complementary *non-drug* therapies based on the resynchronization of the patient's internal rhythms. This model suggests that a therapeutic action is possible by imposing, through harmonic perturbation, an external rhythm on which all internal rhythms will synchronize. In particular, imposing an external circadian rhythm would be very successful insofar as most of the internal (biological or psychological) rhythms are circadian. This *chronotherapy* can be done by acting on the sleep / wake cycle (for example, by partial sleep deprivation or advance of sleep phase), or on the light / dark cycle (using light therapy for depressive phases and darkness for manic phases). This way of healing patients suffering from bipolar disorder, which has already been successfully tested (Wu et al. 2009), finds here its theoretical justification. This rudimentary model of bipolar disorder could of course be refined by integrating its clinical features in more detail. However, even in this ideal form it can provide a consistent explanation of the mood shifts and of their acceleration, it allows to make predictions on the average duration of episodes and suggests new non-drug ways of healing.

5.5 Distant Healing

"Distant healing" denotes a healing process for which the healer and the sick person are not located next to each other. Distant healing must be regarded as a form of *spiritual healing*, with the term "spiritual" being generally interpreted in two ways. It first refers to the healer's *intentional* act capable of healing a person without conventional interventions, that is, without using drugs or other chemical or physical means. It also refers to the fact that such healing is possible because the healer reaches a *higher dimension of consciousness*, connecting with a Higher Self, God or another deity, thus revealing an extraordinary potential of empathy, love, creativity and meaning (Dossey 1993). In all cultures, there have been many occurrences of spiritual healing reported, which emphasize the healers' efficacy of the caring intention. For example, Szekely (1975) reports that the *Essenes, a* Jewish brotherhood close to the Qumran community where Jesus of Nazareth studied, used their hands or precious stones as a tool for transmitting their "healing energy" to heal present or remote people. King (1996) reports that in the ancient Hawaiian tradition Polynesians developed holistic techniques involving the emotional and spiritual aspects of the human to prevent and cure illness. Spiritual healing has indeed been practiced according to several methods, by prayer but also by using laying on of hands, by medical dowsing or other rituals, whose specific strengths and weaknesses have been classified and analyzed by Daniel J. Benor (2001, p. 139–149). This author reports a very large number of experiments carried out, for the most part, according to the rigorous criteria of experimental science and which significantly demonstrate the effectiveness of spiritual healing. In 2001, Benor reported that, out of 191 controlled experiments of spiritual healing performed according to scientific methodology (that is, using blind or double-blind, randomized conditions), two third of them show statistically significant results. More precisely, 83 of them show effects for which the probability that they occur by chance is less than 0.01 (that is, $p = 0.01$), while 41 others show effects for which the probability they occur by chance is less than 0.02 or 0.05 ($p = 0.02$ or 0.05), their methodology being also analyzed according to strict criteria of quality (Benor 2001, p.189–190 and 371–376). The search for evidence corroborating the reality of distant healing continues and the reasons for its rejection by mainstream science as well as its possible explanations within its framework are discussed in detail in the works of Dossey and Radin (Dossey 2008; Radin et al. 2015).

As mentioned above, spiritual healing does not require the healer to be in the presence of the sick person, which means that it can take the form of *distant* healing. Regardless of any spatial distance, a distant healing intention can be defined as "a compassionate mental act directed toward the health and wellbeing of a distant person" (Radin et al. 2015). Distant healing uses rituals, such as in shamanic traditions (Hultkrantz 1997) or female rituals of power within antique Egyptian magic (Mc Ginnis 2012), and it can also be realized through intercessory prayer on which we will mostly focus here. Prayer involves a mental activity, as well as speech. From a religious point of view, prayer is understood as an elevation of the spirit towards God to either thank him or to submit him a request. Along this line of thought, prayer has been described as "the practice of the presence of God" by the sixteenth century Carmelite monk Brother Lawrence (Lawrence 1692). *Intercessory* prayer is the act of praying on behalf of others and it is used for spiritual healing, particularly distant healing. The efficacy of intercessory prayer for healing has been tested in many experiments, which have been extensively reported by Dossey (1993) and by Benor (2001). In this last section we will first mention a few significant avenues in the research on the phenomenon of distant healing, referring to experimental studies. This research is closely related with the more general study of the possibility of communication between two persons who are physically and informationally isolated from each other (Tressoldi et al. 2011; Tressoldi and Khrennikov 2012; Rhine 1964) or with that of a possible mental effect on a remote object (Radin and Nelson 1989; Schmidt 1974, 1976), which belong to the field of parapsychology. However, for our purpose we will concentrate on *the healing effects* of such a non-local communication, which fundamentally involves the healer's caring intention and *emotional state*. We will then explain that the possibility of distant healing is a consequence of mind-body entanglement and propose an appropriate, quantum-like model of it.

5.5.1 A Few Experimental Studies on Distant Healing Phenomena

The possibility of distant healing has been, and still, is subject of controversy. In one of the first published studies of this phenomenon, Galton (1872) concluded that prayer does not have any efficacy on people's well-being and longevity but that, on the contrary, it is harmful. To arrive at this conclusion, Galton mentioned the fact that sovereign heads of state, like monarchs and high religious authorities generally had the shortest lives although they were daily and abundantly prayed for. However, as mentioned by Dossey (2008), Galton's study is skewed because he fails to take into account several important factors. One of these factors is the essential fact, also pointed out by John Polkinghorne (1989), a physicist and fellow of the royal Society, that sovereigns were more often exposed to the ministrations of the medical profession whose danger was great at that time. Other experiments have been cited by skeptics in order to dismiss the reality of spiritual healing, such as the one reported by Collipp (1969) aiming to assess the power of intercessory prayer for healing children with leukemia. Its flaws, commented by Dossey (1993, p. 176–177), regard several essential conditions such as the small size of the patient population or the failure to take into account the possible difference in the severity of the disease in the two (prayer-for and control) groups.

However, much more rigorous studies have shown significant distant healing "effects". Dossey (2008) reported that in June 2008, around 22 major control-studies of distant healing using intercessory prayer were published and that half of them have demonstrated statistically significant results. Some of the most important ones will be reported below. More extensively, Jonas and Crawford found 2200 published reports on spiritual healing and similar phenomena, including 122 laboratory studies and 80 randomized controlled trials (Jonas and Crawford 2003). These studies are not only about intentional effects on humans but also on other non-human biological systems: at the tissue level, involving for example red blood cells (Braud 1990) or cancer cells (Bengston and Krinsley 2000), at the microbial level, about the growth or mutation of bacteria (Nash 1982, 1984) and that of fungi (Barry 1968), or at the molecular level, about the motility of simple organisms like paramecia or the enzyme kinetics (Richmond 1952). The latter researches on the efficacy of intercessory prayer for well-being and healing of non-human beings, which actually have even more significant results (Benor 2001, Chap. 4), are important because they overcome a fundamental difficulty in studying the efficacy of intercessory prayer in humans. This difficulty is that the results could be completely distorted by the possibility that other people who do not belong to the designated "prayer group" of the experiment, like family or loved ones, may also pray for the sick person, or even that the sick person may pray for themselves. The study of the efficacy of intercessory prayer in non-human subjects thus allows to overcome this possible flaw of the study for humans. The studies mentioned by Jonas and Crawford also dealt with the effect of human intention on inanimate systems, which is relevant to psychokinesis insofar as they regard the possibility of remote influence of humans on physical systems, such as random generators, without interacting with them (Radin and Nelson 1989; Schmidt 1974). The latter psychokinetic effects could shed some light on the phenomenon of distant healing by interpreting the connection established between the healer and the person to heal as a case of psychokinetic phenomenon where the healer's mind "acts" on a distant material body. Note that Jonas and Crawford introduced a label of quality by applying CONSORT criteria[20] to these studies, giving on an A to E scale the highest grade (A) to most of the mind-matter psychokinetic studies and grade B to those about the effects of prayer on healing.

Returning now to healing intention therapies of humans, let us first consider Byrd's study (1988), which played an important role because it showed that the phenomenon of distant healing can be tested *experimentally* using scientific methodology, based on statistical studies. This study involved about 400 patients with acute heart problems, hospitalized in the heart care unit of San Francisco General Hospital. Several prayer groups disseminated across the country were asked to pray for only half of these patients. Comparing the evolution of disease for this prayed-for group of patients with that of the control group has shown a 5–7% advantage for the prayed-for group of patients regarding the need for medical treatment (antibiotic or diuretic) and the possible complications of the disease (pneumonia, cardiopulmonary arrests, angina pectoris), a percentage which is not very important, but nevertheless statistically significant. However, most significant is the fact that *no* patient in the prayed-for group required cardiopulmonary resuscitation while 12 patients in the control group needed it. Byrd's study has been replicated by William Harris, from the Mid-America Heart Institute in Kansas City, who performed a randomized controlled trial of the effects of intercessory prayer at a distance for 990 ninety patients admitted to the coronary care unit (Harris et al. 1999). These patients were randomized to receive or not intercessory prayer from prayer groups who prayed for them daily for 4 weeks. The patients were unaware of this and the intercessors did not know them and have never met them; they only knew the first names of the patients they prayed for. This study showed a significant statistical difference in the medical course from the admission to hospital discharge based on 35 (weighted or unweighted) variables between the 466 patients who received intercessory prayer and the 524 patients of the usual care group. The mean ± SEM (standard error of the mean) weighted course score was 6.35 ± 0.26 for the prayer group vs 7.13 ± 0.27 ($p = 0.04$) for the usual care group; and, respectively, 2.7 ± 0.1 vs 3.0 ± 0.1 ($p = 0.04$) for the mean unweighted course score.

Let us also mention two important randomized studies, still involving coronary patients, whose mixed results have given rise to discussion. The first one has been directed by Krucoff and team (2001), at Duke University Medical Center, on coronary patients whose course of the disease was followed by mean of a cardiac catheterization. This study showed a significant difference (50–100%) between the

[20] CONSORT means "Consolidated Standards for Reporting Trials". It is a set of quality recommendations for reporting randomised trials.

patients who unknowingly received prayer and those of the control group. This study was then replicated in 8 more major hospitals throughout the country, on 748 coronary patients and by appealing to a variety of prayer groups. The global data did not reveal significant advantage for the prayed-for group but a finer data analysis nevertheless showed that prayer has significant effects on patients practicing other spiritual therapies, like music, imagery or touch therapy (Dossey 2008). The second study, called STEP (for Study of the Therapeutic Effects of Intercessory Prayer), has provided negative results and has often been cited for discrediting any possible healing effect of prayer. This study, conducted by Benson and team at Harvard Medical School (Benson et al. 2006), aimed to assess the role of uncertainty (of being prayed for or not) on the healing effects of prayer. It involved 1802 patients undergoing coronary-artery bypass surgery in six US hospitals and divided into three groups of equal population. Group I patients were prayed for but were not sure whether or not they had been prayed for; group II patients were not prayed for and also did not know whether or not they were prayed for; group III patients were sure they would be prayed for and were in fact prayed for. This study showed that, contrary to the expected results, the patients of group III, who were rightly informed they would be prayed for, clearly had the worst score regarding the postoperative complication. Several factors could explain the negative results of Benson's study, like the possible influence of extraneous prayer (received for example from friends and family), possible errors due to false randomizations (group III more predisposed to postoperative complication) or possible reactions of groups I and II patients to soliciting extraneous prayer or experimenters' effects. However, as emphasized by Dossey (2008), the fundamental flaw in Benson's study and in all other studies using blind randomization is that *they completely neglect the essential role played by the intertwinement between the healer and the person to heal.* The problem with such "pure" statistical methodologies, which are free of all subjective elements, is that they seem to miss the essential point on which are *intentional* healing effects based, namely *the emotional connection* between the healer and the person to heal. This essential, yet obvious point must be taken into account in any attempt to model distant healing.

To end this brief survey of distant healing studies, let us mention that, while the previous reported studies deal with *spatial* distance, the possibility of distant healing *in time* has also been explored, even this is to a lesser degree. "Distant healing in time" means that the healer and the person to heal are located at two different moments in time. For example, Leibovici (2001) conducted a double blind, randomized controlled trial on retroactive intervention involving 3393 adult patients with bloodstream infections detected and healed at the Rabin Medical Center, Beilinson Campus, Israel, during 1990–96. The trial began in July 2000 when these patients were randomized into a control group and an intervention group to which was assigned a remote, retroactive intercessory prayer for their well-being and full recovery. The measured variables were mortality, length of stay in hospital and duration of fever. This statistical study showed that for the patients (during years 1990–96) who received retroactive intercessory prayer (performed in year 2000) the length of stay in hospital (whose maximum was 165 days) was significantly shorter than that of the control group that did not receive the retroactive prayer (its maximum being 320 days), the probability that this difference has occurred by chance being $p = 0.01$, and a fewer mortality, although less significant (28.1% in the intervention group and 30.2% in the control group, with $p = 0.4$). If this result seems strange with regard to the conventional concepts of time and causation, or even completely flawed regarding its methodology, it has been corroborated by several significant statistical studies, which have evaluated a retroactive influence of some seemingly unpredictable future events on an individual's cognitive, affective or physiological activity. For example, a meta-analysis of 26 reports published between 1978 and 2010 conducted by Mossbridge, Tressoldi and Utts (2012) revealed a physiological anticipation of seemingly unpredictable stimuli with a significant overall effect size 0.21, within 95% confidence interval 0.13–0.29. The physiological anticipation has been assessed by measuring the change in several physiological variables, such as electrodermal activity or heart rate, while stimuli consisted for example in a randomly intermixed series of violent and emotionally neutral photographs which could not be predicted. Mossbridge and team's meta-analysis has been replicated by Duggan and Tressoldi (2018) who completed it with 19 new peer and non-peer reviewed studies performed from January 2008 to June 2018, which confirmed (and even improved) its results.[21]

5.5.2 How Does Distant Healing Work?

As noticed by Levin (1996), the answer to this question can be of two kinds. Either we think that the phenomenon of distant healing has a naturalistic origin and can be explained within the framework of present or future science, or we think that this phenomenon has a supernatural origin and transcends scientific knowledge. In the latter case this "explanation" appeals to the healing power of God or a deity acting beyond the laws of the natural world and distant healing is thus conceived as an act of divine blessing. These two kinds of explanations seem to rely, respectively, on two incompatible, scientific-like and religious-like, readings of the world that some have nevertheless tried to reconcile (see below). Claiming that distant healing has an ultimate supernatural (divine) origin, that it exclusively relies on religious faith and that it transcends the laws of nature can easily be challenged. As mentioned in Benor's extensive report of healers' experiences (Benor 2001, Chap. 1), many spiritual healings, whether initiated remotely or not, have been successful even though healers were not particularly prone to religious beliefs. Namely, for most healers, such successful "effects" have often been obtained by laying on of hands, as is the case of Olga Worral reported by Cerutti (1975), or by practicing "bioenergy" field manipulations, as is the case of the Krivorotovs (Benor 2001, p. 69–75), these

[21] These studies showed an overall weighted effect size 0.28, estimated with a frequentist multilevel random model, with a 95% confidence interval 0.18–0.38.

techniques being designed and practiced *independently from any religious belief*. Moreover, if pushed to the extreme, the religious interpretation of the efficacy of intercessory prayer for healing seems difficult to accept for the following reasons. As emphasized by Dawkins (1997), this kind of claim quickly turns into an argument of authority forcing us to unconditionally accept revelation and faith instead of scientific evidence. This claim suggests that we would have nothing more to say about distant healing or about any phenomenon not yet explained by science because it is decreed by a religious authority that this phenomenon transcends the scientific approach on *principle*. Moreover, this line of thought may lead to the rejection of the scientific studies about the efficacy of intercessory prayer on health by questioning the religious faith of the prayer group involved in these experiments or by questioning their lack of any particular religious affiliation (since healers can a priori be Jews, Christians, Muslims, Buddhists or atheists). In other words, this religious position leads to the idea that the phenomenon of distant healing is reserved to an elite, namely to some believers chosen by God and capable of welcoming divine grace, instead of being usable as a medical tool for *any* ill person. This extreme religious position about distant healing phenomena thus opposes its scientific study and its use for medical purpose.

Actually, many famous religious people have suggested that the method and results of science are quite compatible with religious faith by claiming that any natural force or natural power, namely here the sick person's life energy, *originates from God* (Wilson 1972). Levin (1996) emphasizes that the natural effects of God's supernatural power and therefore the absence of conflict between modern science and Jewish religious texts,[22] have been clearly stated in prominent theological treatises, as in the works of Maimonides (Bleich 1993) and Naftali Berg (Berg 1994). Along this line of thought, one can also mention Jesuit priest, paleontologist and philosopher Teilhard de Chardin (1955), whose religious conception explicitly integrates natural selection and the notion of random genetic mutations. Such an attempt to reconcile religious and scientific readings of the world thus rely on the idea that all phenomena, whether classified as "natural" and knowable by science or "supernatural" and transcending science, ultimately originate in God. However, whether we think that the phenomenon of distant healing has something to do with the healer's religious beliefs or not, the key point remains to improve our knowledge of it and the best way to do this seems to require a rigorous scientific methodology. We will then concentrate on the possible *naturalistic* approaches to distant healing.

The existence of a mind-matter interaction has been proposed by many authors for explaining the efficacy of distant healing. Despite of the apparent physical and informational isolation of the healer and the person to heal, there would nevertheless exist a (very subtle) "force", "energy" or "field", still unknown by contemporary physics and biology, propagating *through space* from the healer to the sick person and thus "carrying", in a sense that needs to be clarified, the healer's healing

[22] In particular, Torah's affirmation that God transcends the physical laws of the world which He created.

intention. These “latent forces or energies” are qualified by Levin and Vanderpool (1989) as “superempirical” to emphasize that “the existence of functioning of such energies have not yet been empirically verified to the consensus satisfaction of mainstream science” (Levin 1996). As reported by Benor (2001, Chap 1), many healers have suggested the existence of such an “energy that can restore health” (Mc Manaway and Turcan 1983). This energy would take birth in the healer and spreads towards the person in need of healing, through hands or through some vehicle capable of storing it (Sherman 1972). Dean Kraft, a healer who is said to have psychokinetic powers, asserted that (Kraft 1981):

> … something like energy (in my mind it was light) was coming out of me and connecting to the object, and a similar model seems to apply to healing. Indeed, almost all the people I work with claimed to feel a pulsating energy, like electricity, passing into their bodies, often accompanied by a sensation of warmth.

Along this line of thought, let us recall the recent interactionist interpretation of such a mind-matter connection within a quantum framework that we mentioned in Chap. 1. This approach is grounded on the subjective interpretation of the collapse of the wave function according to which the subject’s “consciousness”, “mind” or “faculty of introspection” would play an essential role in the sudden transition of the wave function from a state of superposition to the unique state observed. For example, Walker has suggested to quantify the action of mind over matter using the amount of Shannon’s statistical information associated with the probability of this transition (Walker 1975), a model which has been refined by Mattuck (1980) by assuming that mind would be capable of performing a quantum transition in several steps. However, as noticed in Chap. 1 (Sect. 1.3), and as shown experimentally (Sect. 4.4), it seems that no consistent interactionist interpretation of the mind-body connection has been provided and, therefore, this negative conclusion also holds for the latter quantum-like models. Also note that, even under the assumption that such an “interaction” would take place between the mental and physical *properties* of a unique material substance, rejecting thus Descartes’ dualism of substances, such an idea of mind-matter “interaction” still seems inconsistent. Kim’s exclusion argument against non-reductionist physicalism, also presented in Chap. 1, can still be applied here: physicalism is grounded on a principle of causal closure of the material world, which excludes the causal efficacy of any non-physical event, such as a thought or a mental state, on the material world. Consequently, it seems that distant healing cannot be explained by appealing to a concept of “causal interaction”, whatever its nature, between the healer and the sick person.

5.5.3 Distant Healing and Generalized Entanglement

We will now propose a quantum-like model of distant healing grounded on the concept of generalized entanglement, which underpins our systemic approach of the mind-body connection. By explicitly referring to the non-local connection between the healer's emotional state and that of the person to heal, this model implements Dossey's concept of "non-local mind" (Dossey 2015) according to which.

> ".... consciousness cannot be completely localized or confined to specific points in space, such as brains and bodies, or to discrete points in time such as at the present moment."

It also strongly echoes with Radin's idea of "entangled minds" (Radin 2006) that explicitly refers to the quantum concept of entanglement. However, we do not conceive here this entanglement of minds as a specific application of the purely *physical* concept of quantum entanglement -for example by appealing to the physical entanglement of electromagnetic fields emitted by our bodies (Radin 2007). As explained in this book, the key concept on which is based our approach to the mind-body connection is that of *generalized* entanglement, which, although formally similar to Schrödinger's concept of quantum entanglement, has been defined (in Chap. 4) *to be applied to generalized systems, which are not necessarily physical.* The concept of quantum entanglement only concerns *physical* systems, whether microscopic or not, whose behaviour is governed by the laws of *quantum physics,* which is not the case of minds (not brains) that are not physical systems. Our approach to distant healing phenomena is also close to Chalmers' extended mind thesis (Clark and Chalmers 1998) and to other lines of thought relying on the concept of an undivided universe endowed with a collective memory, such as Bohm's implicate order theory (Bohm and Hiley 1993) and Sheldrake's concept of "morphic resonance" (Sheldrake 1995).

As is the case for the mind-body connection, the phenomenon of distant healing can be explained in terms of generalized entanglement. The key idea is that when the healer (h) and the person to heal (h′), are *emotionally entangled*, the emotional observables of h are also entangled with the somatic observables of h′, due to the entanglement of their own emotional and bodily observables. To say that two persons are "emotionally entangled" means that their respective emotional observables are entangled.[23] The phenomenon of distant healing can thus be regarded as a kind of *entanglement swapping*, where the entanglement of the emotional and somatic observables of h′ is transferred into the entanglement of the emotional observables of h (the healer) and the somatic observables of h′, thanks to the emotional entanglement between h and h′. Within the quantum-like framework presented in Chap. 4, this process of entanglement swapping can be written as follows[24]:

Before emotional entanglement of h and h′, their entangled psychosomatic states can be written as follows in the respective basis $| M_a, P_b >$ and $| M_{a'}, P_{b'} >$ of the psychophysical spaces ***H*** and ***H′:***

- Entangled psychosomatic state of h: $| \Xi >_h = \Sigma_{a,b}\, \alpha_h\, (a,b) \,| M^h_a, P^h_b>$

[23] Remember that the entanglement of two observables of two subsystems (namely, h and h') means that these observables are strongly correlated regardless of any spatial or temporal separation of h and h' (see Chap. 4).

[24] We use here a synthetic and simplified version of psychosomatic states, as explained in Chap. 4 and already implemented in this chapter (Sects. 5.2–5.4). The letters M and P respectively denote emotional tones (which are made of several interdependent emotions) and biological programs.

- Entangled psychosomatic state of h′: $|\Xi>_{h'} = \Sigma_{a',b'} \alpha_{h}'(a',b')\ |\ M^{h'}_{a'}, P^{h'}_{b'}>$

After emotional entanglement of h and h′, the psychosomatic state of the composite system h + h′ is an entangled state of the product space ***H* ⊗ *H’***:

$$\left|\Xi>_{h+h'} = \Sigma_{c,d}\ \alpha_{h+h'}(c,d)\right|M^{h}_{c}, P^{h}_{d}; M^{h'}_{c}, P^{h'}_{d}>.$$

In this new state, the healer’s psychosomatic state is strongly correlated with the sick’s psychosomatic state regardless or any spatial or temporal distance, which means that if the mental healer’s state is defined by the emotional tonality M^{h}_{c}, the sick’s somatic state is necessarily given by $P^{h'}_{d}$.

Distant healing is thus possible at the two following conditions: (1) *h and h′ must be emotionally entangled*, which leads us to wonder how this can be done, and (2) *h can reach an appropriate emotional state whose properties make possible the healing of h′*. More precisely, this appropriate emotional state reached by the healer is defined by specific values of emotional observables correlated to the somatic observables of the sick person. This condition will lead us to investigate the healers’ appropriate emotional states.

For realizing condition (1) of emotional entanglement between the healer and the patient no physical interaction is of course required, unlike the preparation of entangled systems in the physical domain. Without especially referring to prayer, Walach, Tressoldi and Pederzoli (Walach et al. 2016) have described the general method to get such a mind-mind entanglement between two persons and they also have mentioned the way this entanglement can be assessed (Walach et al. 2016):

> The typical procedure to create mental entanglement consists of asking two people who are sensorially isolated to mentally connect with each other. One of them is then given some information, for example an image or sensory stimulation, which constitutes the variables used to observe the non-local correlation between two partners by measuring the coincidences between what is given to one and what the other reports.

Many more or less significant statistical studies of this typical procedure of mind-mind entanglement or that of a variant of this procedure, where the person who is supposed to share the information is substituted by just a source of information, have been reported. Let us first mention the studies conducted by Rhine (1964) at Duke University, USA, and by Vasiliev (1965) in Russia that already provided a big corpus of data. They demonstrate that some subjects can seemingly obtain information from people without any physical or sensorial connection with them. More recently, a statistical study conducted by Baptista and team (2015), where a subject placed in normal sensorial conditions had to mentally connect with distant information (about a picture, for example) that cannot be received by physical means, provided a correlation coefficient $r = 0.12 \pm 0.02$. We can also mention Tressoldi’s meta-analysis (2011) of 108 experiments where the subject who had to mentally connect with distant information within Ganzfeld sensorial conditions in order to improve the mental signal-to-noise ratio. To be placed in “Ganzfeld sensorial conditions” means that this person is comfortably installed in an armchair with their eyes closed and wearing an audio headset emitting a white noise.[25] This person then describes what they perceive while the second person, located in another room, looks at a randomly selected target (usually an image). This meta-analysis has provided a correlation coefficient of $r = 0.06 \pm 0.02$, which is rather weak.

Beyond the statistical significance and reliability of such studies about the possibility of non-local communication of emotionally neutral information,[26] let us recall that our purpose requires to focus on *the emotional component* of the mental connection between two persons. As noticed by Dossey about the previously mentioned Benson's study, emotional entanglement should provide much more significant experimental results than entanglement with not-emotionally charged information in as much as emotion literally "sets in motion" our mind and our body. Other experimental tests involving more clearly emotions are then required to study the feasibility and the reliability of the process of emotional entanglement between the healer and the patient.

If further data are needed to study such an emotional connection, its potential risks and side-effects can already be mentioned. The later have been observed and studied by Walter von Lucadou at Freiburg's Parapsychological Counselling Office[27] (Lucadou 2014). Lucadou reports that long after treatment patients can still experience a kind of "telepathic contact" with the healer, a connection often described by them as intrusive, even threatening. It is tempting for these patients to interpret such a phenomenon in terms of "hacking": the required non-local, emotional connection between the healer and the patient has been distorted or corrupted by the influence of parasitic, non-expected emotions generated by a third party whose intention can be hostile. However, according to Lucadou these side-effects, called by him the "bewitchment syndrome", mainly result from an unsolved emotional entanglement between the healer and the patient (Lucadou 2003).

Also note that a systemic limitation of such a non-local, emotional connection could be imposed by the "non-transmission axiom", or NT-axiom, which, according to Lucadou, Römer and Walach (2007), should reasonably be set in a generalized version of quantum theory (Lucadou 2019). This axiom asserts that generalized entanglement correlations cannot be used for transmitting signals or controllable influences (Lucadou et al. 2007), which can be regarded as the generalization of the no-communicating theorem (or Eberhard's theorem) of quantum theory asserting that quantum entanglement cannot be used to transfer a signal.[28] In order for the

---

[25] A « white noise » is a background noise generated stochastically so that the same power spectral density is assigned at all frequencies. It is used here to mask or ignore other noises coming from the environment.

[26] Note that the possibility of non-local communication could indeed first be tested experimentally by physics (Uzan 2020) by using more consensually admitted protocols, such as measuring the correlations between photon polarizations.

[27] Freiburg's Parapsychological Counselling Office helps individuals to cope with paranormal experiences adequately. It benefits from government support (Lucadou 2014).

[28] The non-communication theorem shows that in a two-party system no action of one of the parties (i.e., the choice of the observable that it decides to measure) is detectable by the other party by

NT-axiom to be satisfied, any attempt to use a non-local correlation for transmitting a signal would make this connection vanish or change its effect in an unpredictable way (Lucadou et al. 2007). For our purpose this means that the NT-axiom would limit the replicability of such healing action and give rise to decline and displacement effects, which could limit the reliability of such a treatment and even be harmful for the patient.

In order to deal with condition (2), let us now investigate the nature of the healer's felt emotions. In the placebo phenomenon examined in Sect. 5.3, reaching a state of optimism and confidence in the physician and the prescribed treatment makes the triggering of bodily healing processes possible. However, in the case of distant healing by intercessory prayer or the other means listed above there is no physician and no drug treatment to trust in. Praying is addressing oneself to God or to another deity, which places the one who prays in a state of trust in Him and devotion, as they ask Him to provide the means to heal the sick, either themselves or a loved one. The one who prays places their trust in God and has complete confidence in His healing power, as they are in a state of total devotion. In other cases of spiritual healing which use rituals, laying on of hands and/or visualization (Kraft 1981), as well as in the placebo phenomenon, this state of belief, confidence and love is experienced by the healer even though there is no address to God or to a religious authority. As noticed by several healers, the efficacy of the different healing methods of spiritual healing seems to rely in an essential way on the healer's ability to concentrate and to be "centered", that is, to clear their mind and relax their body in order to reach an inner state of stability, empathy and union with the sick (Benor 2001 p. 103 and 143–144). Therefore, it seems that the emotional state of optimism, confidence and love, which is needed for healing, is not reached thanks to a particular belief but *by the very capability of believing, concentrating and "centering"*. Similar to Lambert's observation (2018) about miracles, we can say that what seems essential to emphasize in the phenomenon of spiritual healing (and a fortiori in distant healing) is not really the specific healer's belief but *their power to form beliefs*, which has been called in French "crédivité"[29] by Joseph Fusswerk-Fursay (1978). "Crédivité" is the condition of possibility to reach a state of confidence, empathy and love, that is, to generate the required appropriate positive emotions that are strongly correlated with the biological programs involved in the bodily healing processes. It is this power of belief that allows the one who prays to develop empathy and compassion for the sick. If the strength of religious prayer is based on the belief in God and in His healing power, the human's "crédivité" can indeed have a wide range of objects of belief, from religious beliefs to the belief in their own ability to heal (Benor 2001, Chap. 1). Accordingly, spiritual healing can be performed through

making measurements. However, this theorem is based on a statistical description of the systems under consideration, which cannot be necessarily be provided in a weakened version of generalized quantum theory. It is thus set as an axiom.

[29] The French neologism "crédivité", which designates the human faculty of believing, must be distinguished from "belief", which is *its object*.

prayer or by practicing rituals, such as laying on of hands, visualization (Kraft 1981) and therapeutic touch (Leshan and Margeneau 1980).

Thus, thanks to the emotional connection between the healer and the sick the entanglement of emotional and somatic observables of the latter is transferred into the entanglement of their somatic observables and the emotional observables of the healer. This phenomenon of entanglement swapping makes possible the triggering of healing processes in the sick's body, regardless of any spatial or temporal distance between them, provided that the healer has reached a "positive" emotional state of love, optimism and confidence.

Chapter 6
Conclusion

An exploration of the difficulties of the yet intuitive causal interpretation of the mind-body connection based on logico-philosophical arguments and experimental results led us to suggest a radical change of perspective to deal with the nature of this connection. Instead of trying to connect as best we can mental life and the functioning of the body, either by identifying them and leaving then aside subjective experience or by appealing to an elusive causal relationship, we started from the notion of psychosomatic unity of the individual and looked for its conditions of possibility. We have shown in Chap. 3 that what makes this unity possible is the *generalized entanglement relationship* that connects a person's subjective experience and the functioning of their body in a specific way.

To see this, we have highlighted the properties of the mind-body connection by particularly focusing on emotions that sit at the mind-body interface and express themselves through both bodily changes and subjective experience. We have shown that, in addition to the properties of inseparability and specificity that have been emphasized by psychosomaticians, psychophysical correlations satisfy two other fundamental properties, namely, those of a-causality and non-locality, which characterize the relation of generalized entanglement. We have concluded that the psychosomatic unity of the individual *requires* the mind-body connection to be a relation of generalized entanglement. As explained in Chap. 4, the concept of mind-body entanglement finds a fruitful mathematical representation within a general systems theory obtained by generalizing the key-concepts of quantum theory in order to apply them to all kinds of systems and observables, while keeping its full mathematical structure (non-commutative C*-algebra of operators and Hilbert space formalism). Moreover, its validity can be assessed experimentally by applying a Bell-like statistical test on pairs of complementary observables respectively defined in the mental and somatic domains.

We have shown in Chap. 5 that this mathematical framework is a valuable tool for exploring the radical consequences of this approach in the medical field. The concept of mind-body entanglement can for example explain the existence of a

P. Uzan, *Mind-Body Entanglement*, SpringerBriefs in Philosophy,
https://doi.org/10.1007/978-3-030-90173-8_6

"body memory" and leads to a better understanding of the genesis and evolution of internal diseases, then suggesting the development of mind-body therapies that could effectively supplement or even replace drug therapies. It provides new insights into mental disorders, their classification and their relationship with brain functioning, and therefore suggests alternative or complementary approaches for their treatment. Furthermore, it sets the theoretical basis for tackling some paradigmatic self-healing methods, such as the placebo "effect", biofeedback, and meditation, which remain totally enigmatic according to the local-causal way of thinking. It also provides an explanation of distant healing in terms of entanglement swapping.

All these alternative or complementary approaches to illness and healing conceive illness in terms of psychosomatic imbalance that the mentioned practices make it possible to restore. The efficacy of these "non-conventional" practices, those of oriental medicine or those utilizing the placebo phenomenon, biofeedback, resynchronization of internal rhythms, meditation or spiritual strength, is due to the fact that *they directly refer to the genuine nature of the mind-body connection, namely to the generalized entanglement relationship between thoughts and body functioning, between emotional and somatic observables.* As shown in detail in Chap. 5 for *all* the models presented in the field of health and care, the latter have been developed *within* the framework of the systemic approach to the body-mind problem presented in Chap. 3 and whose mathematical framework is provided in Chap. 4. These alternative healing models are thus *fundamental consequences* of the systemic approach to the mind-body connection in terms of generalized entanglement. They have been built by generalizing the quantum-like concepts of complementarity and entanglement and by applying them to the psychosomatic unity of the individual. They cannot then be developed along the local-causal ontological approach to illness according to which its origin would lie in objectified, well defined entities whose dysfunction would be causally responsible for its development. In contrast with this mainstream concept of illness according to which healing would be reduced to external interventions, such as administration of drugs, surgery or vaccine, the holistic approach to healing presented here, which closely relies on the entanglement nature of the mind-body connection, directly refers to the healing power of the human being.

Consequently, in addition to constituting a significant contribution to the long-standing philosophical debate about the nature of the mind-body connection, this change of perspective based on the concept of generalized entanglement thus allows the development of a very promising approach to illness and healing.

Correction to: A Quantum-like Representation of the Psychosomatic Unity

Correction to:
Chapter 4 in: P. Uzan, *Mind-Body Entanglement*, SpringerBriefs in Philosophy, https://doi.org/10.1007/978-3-030-90173-8_4

Owing to a typographical error on page 80, the formula in the section 4.3.1 has been presented wrongly as:

(Df) R = S(aa,) + S(ab,) + S(b·a,) – S(bb,),

where S (aa,)

The formula has now been updated to

$$(\mathrm{Df})\mathrm{R} = \mathrm{S}(\mathrm{aa}') + \mathrm{S}(\mathrm{ab}') + \mathrm{S}(\mathrm{ba}') - \mathrm{S}(\mathrm{bb}'),$$

where S (aa′)

The updated version of this chapter can be found at
https://doi.org/10.1007/978-3-030-90173-8_4

P. Uzan, *Mind-Body Entanglement*, SpringerBriefs in Philosophy,
https://doi.org/10.1007/978-3-030-90173-8_7

Appendix 1: About the Mathematical Formalism of Quantum Theory

This appendix is an intuitive presentation of the main ideas underlying quantum theory: C*-algebras of operators are utilized to represent the quantities that can be measured on a system and the operations on these quantities, while the Hilbert space formalism is utilized to represent its possible states and their temporal evolution. The main idea underlying the representation of a composite system will also be mentioned.

C* Algebra of Observables

Generally speaking, the measurement of a quantity, called "observable", on a system is conceived as an operation carried out on the space of possible states of this system. In classical mechanics, these operations are functions of the generalized coordinates of the point which are encoded in the state of the system in the phase space[1] and these functions compute the (unique) pre-determined values of the measured quantities. However, in quantum mechanics the value taken by a measurable quantity is not totally determined by the state of the system on which is performed this measurement. Quantum theory only provides us, via Born's rule, with the statistical distribution of its possible values. In such a probabilistic theory of prevision whose structure has been studied by Pauline Février and Jean Louis Destouches (Février et al. 1994), a measurable quantity cannot be represented, like in classical mechanics, by a function but it can be very conveniently represented by a hermitean operator acting on the state space and whose eigenvalues, which are real numbers, provide the statistical distribution of its possible values (Hughes 1989; Bitbol 1996, Chap. 2). The set

[1]The state space of a system composed of N material points is, in classical mechanics, a 6 N-dimensional space (3 dimensions for the three coordinates of position and 3 dimensions for the three coordinates of momentum).

P. Uzan, *Mind-Body Entanglement*, SpringerBriefs in Philosophy,
https://doi.org/10.1007/978-3-030-90173-8

of such operators-observables endowed with the usual operations (combination, addition and multiplication by a complex number) and the operation *, called involution, form a C*-algebra built on complex numbers if these operations satisfy the following properties, for any operators A and B[2]:

$A^{**} = A$
$(A + B)^* = A^* + B^*$
$(AB)^* = B^*A^*$
$(u\,A)^* = \bar{u}\,A^*$, where u is a complex number and ū its complex conjugate.
$\|A\,A^*\| = \|A\|^2$

The order effect in the successive measurement of two observables, for example in the successive measurement of two spins in two different directions (Cohen-Tannoudji et al. 1973, pp. 231–235), is captured by the *non-commutativity* of the C*-algebra associated to the system under consideration. In quantum theory and in its generalized version used in this book the observables *do not a priori commute*, which means that the order in which the mental or somatic variables are measured must be taken into account. The underlying algebraic structure of the mind-body entanglement model is thus a non-commutative C*-algebra. Non-commutative C* algebra also allow capturing the concept of entanglement by referring to couples of complementary observables respectively defined on the two parts of a system. This essential point is clarified in Sect. 4.3, by appealing to recent results that extend Bell and Tsirelson inequalities.

The Hilbert Space Formalism

A Hilbert space is a vector space endowed with an inner product and which is complete regarding the convergence of series (a property which is indeed significant only for infinite Hilbert spaces).

It can be shown that any C*-algebra is isomorphic to a closed *-algebra of the bounded operators B(H) of a Hilbert space H, the involution * being in correspondence with the operation of taking the adjoint sub-space for any sub-space of B(H). Thus, a complex Hilbert space can be associated to any system whose observables form a C*-algebra, its states being represented by vectors (and not by points of the phase space, like in classical mechanics). The components of a vector-state onto a basis of H are square-summable functions (as is the case for the wave function of position, for example).

The possible results of the measurement are the eigenvalues of the measured observable and the associated probabilities are given by the square of the norm of its components (which are probability amplitudes) in the system's vector-state. This

[2]A more detailed presentation of C*-algebras can be found, for example, in Averson's book (Arveson 1976).

probability measure (in the mathematical sense) is uniquely defined on the subspaces of a Hilbert space, as shown by Gleason's theorem. The measurement of an observable on a system in a given state is mathematically expressed by the projection of this state on the eigenspace associated to the result. Note that the use of *complex* Hilbert spaces is very convenient for defining a single mathematical entity, namely the vector-state, from which can be computed the statistical distribution of the results of *all* possible measurement operations, even in the case of non-commutative observables (Hughes 1989).

The temporal evolution of a quantum system is governed by the Schrödinger equation, which expresses the action of a unitary operator on the system's state. This unitary evolution operator U is such that $U U^* = U^* \cdot U = Id$, where U^* is the adjoint of U, and it preserves the inner product and then the norm. If the Hamiltonian operator H, which plays the role of generator of evolution, is a hermitean operator independent of time, U can be written as: $U = \exp(-iHt)$.

Composite Systems

The state space of a composite system $S = S1 + S2$ is the tensor product of the Hilbert spaces H1 and H2 associated with the sub-systems S1 and S2. The reason is that the square summable functions of $n + m$ variables that are supposed to represent the state of the composite system S (n and m being, respectively, the dimension of H1 and H2) cannot generally be defined as the product of a function on H1 and a function on H2. Its definition requires to be done in the tensor product $H = H1 \otimes H2$, where the linear operation $\otimes$ makes the construction of a basis of H possible from a basis of H1 and a basis of H2. (see also Cohen-Tannoudji et al. 1973, p. 162–163).

Appendix 2: Complementarity of Anger and Disgust

To show the complementarity of the emotions Anger (A) and Disgust (D) we have to compute the commutator of their associated observables and to show that it is different from zero. These observables are, respectively, the projectors $\mathbf{A}_{op} = |A > < A|$ and $\mathbf{D}_{op} = |D > < D|$. The following diagram of Chap. 4 shows the basis {|A>, |¬ A>} and {|D>, |¬ D>} in the same 2-dimensional emotional state space:

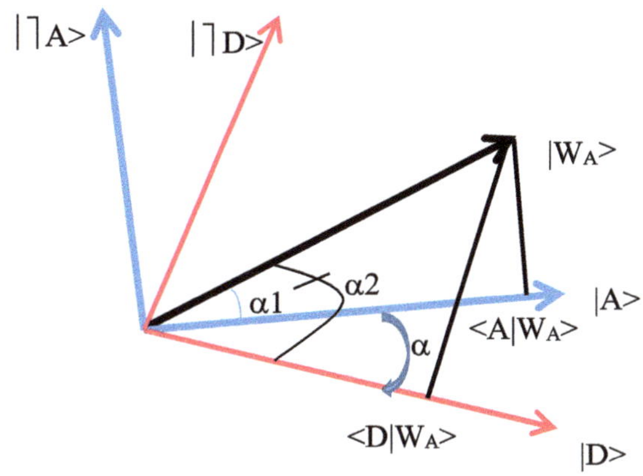

The probability amplitude <D|A > of the transition from the state of pure anger to the state of pure disgust can be computed as follows, from the third line of Table 4.1 (Sect. 4.2, p. 62), keeping the precision to two decimal places given by these experimental data:

$$\cos\alpha 1 = \langle A \mid W_A \rangle = \left(p\left(A / W_A\right)\right)^{1/2} = (5.15 / 7.56)^{1/2} = 0.825, \text{hence } \alpha 1 = 38.2°$$

$$\cos\alpha 2 = \langle D \mid W_A \rangle = \left(p\left(D / W_A\right)\right)^{½} = (1.21 / 7.56)^{1/2} = 0.4, \text{hence } \alpha 2 = 73.8°$$

$$\alpha = \alpha 2 - \alpha 1 = 35.6^{\circ}, \text{and then } \langle A \mid D \rangle = \cos\alpha = 0.847.$$

P. Uzan, *Mind-Body Entanglement*, SpringerBriefs in Philosophy,
https://doi.org/10.1007/978-3-030-90173-8

The state $|D>$ can be written in the A-basis as:

$$|D>=\langle A|D\rangle|A>+<\urcorner A|D>|\urcorner A>,$$

where the probability amplitude $<\urcorner A|D>$ can be computed from $<A|D>$ as follows:

$$<\urcorner A|D>=\left(1-|<A|D>|^2\right)^{1/2}=0.531.$$

Consequently, $|D>=0.847\,|A>+0.531\,|\urcorner A>$.

To compute the commutator of the observables A_{op} and D_{op}, we express their associated matrix in the same orthonormal basis $\{|A>, |\urcorner A>\}$ of the two-dimensional state space. In this basis, the projector $A_{op}=|A><A|$ is associated with the matrix:

$$M(A_{op})=\begin{bmatrix}1 & 0\\ 0 & 0\end{bmatrix}$$

To find the matrix associated with the observable $\mathbf{D}_{op}$ in the basis $\{|A>, |\urcorner A>\}$ we use the aforementioned decomposition of $|D>$ in the basis $\{|A>, |\urcorner A>\}$ to compute the outer product $\mathbf{D}_{op}=|D><D|$ in this basis:

$$M(\mathbf{D}_{op})=(0.847; 0.531)\otimes(0.847; 0.531)=\begin{bmatrix}0.717 & 0.449\\ 0.449 & 0.282\end{bmatrix}$$

The commutator of the observables $\mathbf{A}_{op}$ and $\mathbf{D}_{op}$ can then be computed as:

$$[\mathbf{A}_{op}, \mathbf{D}_{op}]=\begin{bmatrix}1 & 0\\ 0 & 0\end{bmatrix}\begin{bmatrix}0.717 & 0.449\\ 0.449 & 0.282\end{bmatrix}-\begin{bmatrix}0.717 & 0.449\\ 0.449 & 0.282\end{bmatrix}\begin{bmatrix}1 & 0\\ 0 & 0\end{bmatrix}$$
$$=(0.449)\begin{bmatrix}0 & 1\\ -1 & 0\end{bmatrix}.$$

This result shows the *complementarity of the observables* $\boldsymbol{A}_{op}$ *and* $\boldsymbol{D}_{op}$ that, respectively, measure the average intensity of anger and disgust experienced by a subject conditioned in a state of induced anger. Their degree of incompatibility can be estimated by the norm of this commutator:

$$\left\|\left[\mathbf{A}_{op}, \mathbf{D}_{op}\right]\right\|=0.449.$$

Appendix 3: Complementarity of Systolic Pressure and Stroke Volume

For a somatic state | W >, the functions <v | W > and < p | W > can be written, in the v-representation and the p-representation respectively, as the following Fourier transforms:

$$\langle \mathrm{v} | \mathrm{W} \rangle = (\varepsilon)^{-1/2} \int_{\mathrm{p}} \langle \mathrm{p} | \mathrm{W} \rangle \exp 2\pi \mathrm{i} / \varepsilon (\mathrm{p}.\Delta v) \mathrm{dp},$$

$$\langle \mathrm{p} | \mathrm{W} \rangle = (\varepsilon)^{-1/2} \int_{\mathrm{v}} \langle \mathrm{v} | \mathrm{W} \rangle \exp - 2\pi \mathrm{i} / \varepsilon (\mathrm{p}.\Delta v) \mathrm{dv},$$

where $\varepsilon = \Delta U / 1.14$.

Taking now into account, from the expression of <v | W > above, that <v|p > can be computed as:

$$\langle \mathrm{v} | \mathrm{p} \rangle = (\varepsilon)^{-1/2} \exp.\ 2\pi \mathrm{i} / \varepsilon (\mathrm{p}.\Delta v),$$

the action of the observable $\mathbf{P}_{op}$ in the v-representation can be computed as:

$$\langle \mathrm{v} | \mathbf{P}_{\mathrm{op}} | \mathrm{W} \rangle = \int_{\mathrm{p}} \langle \mathrm{v} | \mathrm{p} \rangle \langle \mathrm{p} | \mathbf{P}_{\mathrm{op}} | \mathrm{W} \rangle \mathrm{dp}$$

$$= (\varepsilon)^{-1/2} \int_{\mathrm{p}} \mathrm{p} \langle \mathrm{p} | \mathrm{W} \rangle \exp 2\pi \mathrm{i} / \varepsilon (\mathrm{p}.\Delta v) \mathrm{dp}.$$

In the right member of this equation one can recognize (to within a constant factor) the Fourier transform of p < p | W >, which is the derivative of <v | W > regarding the variable v. Consequently:

$$\langle \mathrm{v} | \mathbf{P}_{\mathrm{op}} | \mathrm{W} \rangle = -\mathrm{i} \varepsilon \partial / \partial \mathrm{v} \langle \mathrm{v} | \mathrm{W} \rangle.$$

P. Uzan, *Mind-Body Entanglement*, SpringerBriefs in Philosophy,
https://doi.org/10.1007/978-3-030-90173-8

The volume observable was defined as:

$$\left\langle \mathrm{v} \mid \mathbf{\Delta V}_{\mathrm{op}} \mid \mathrm{W} \right\rangle =_{\mathrm{df}} \left(\mathrm{v} - \mathrm{v}_{\mathrm{ES}} \right) \left\langle \mathrm{v} \mid \mathrm{W} \right\rangle,$$

where v_{ES} is the end-systolic volume of the left ventricle of the heart. The commutator of the stroke volume operator $\mathbf{\Delta V}_{op}$ and the systolic pressure operator $\mathbf{P}_{op}$ can then be computed as:

$$\left[\mathbf{\Delta V}_{\mathrm{op}}, \mathbf{P}_{\mathrm{op}} \right] = \mathrm{i}\varepsilon\, \mathbf{Id},$$

their degree of incompatibility ε being estimated from the maximal mechanical work of the heart in a cycle ΔU, as $\varepsilon = \Delta U/1.14 \cong 6.54 \times 10^{-2}$.

Appendix 4: Joint Measurement of Emotional and Cardiovascular Observables

In order to explain this meta-analysis, let us present the main steps by referring to the data collected by Kassam and Mendes (2013). All the data have been *normalized to unity* in order to be analyzed within the theoretical framework presented in Sect. 4.3. This means that the spectrum of the observables, which is always bounded, is embedded into the continuous interval [−1, +1]. The normalized value of a variable Y is computed from the raw data according to the following formula:

(Norm) $$\hat{Y} = \left(Y - Y_0\right) / \mid Y_{extr} - Y_0 \mid,$$

Y_{extr} and Y_0 being, respectively, the extremal value of Y and its baseline value corresponding to a neutral conditioning. The main steps of this meta-analysis are presented and illustrated hereafter.

1. The starting point is the tables of raw data which display the change of cardiovascular variables for subjects conditioned in specific emotional states. These double entry tables have the following form, drawn from Kassam's and Mendes' article (2013), where for each emotional state (understood as categorial), Anger and Shame here, the mean value of the cardiovascular changes have been computed from the individual responses of the subjects:

Table D.1 Mean changes and extremal values of cardiovascular variables for anger and shame

	A	Sh	Extremal values
<ΔHR>	15.51, SE = 1.34	10.13, SE = 1.4	15.51 (anger)
<ΔPEP>	−11.29, SE = 1.20	−5.51, SE = 1.26	−11.29 (anger)
<ΔCO>	0.96, SE = 0.17	0.36, SE = 0.18	0.96 (anger)

P. Uzan, *Mind-Body Entanglement*, SpringerBriefs in Philosophy,
https://doi.org/10.1007/978-3-030-90173-8

2. From Table D.1, we compute the *normalized values* corresponding to each emotional conditioning (A or Sh) by using formula (Norm) (Table D.2):

Table D.2 Normalized values of cardiovascular variables

	A	Sh
$<\Delta\hat{H}R>$	1	10.13/15.51 = 0.653
$<\Delta\hat{P}\hat{E}P>$	−1	−5.51/11.29 = − 0.488
$<\Delta\hat{C}\hat{O}>$	1	0.36/0.96 = 0.375

For example, the normalized value of $<\Delta HR>_A$, which is the maximal value of the two reported values of HR (in A- and Sh-condition) is $<\Delta\hat{H}R>_A = 1$, while the normalized value of $<\Delta HR>_{Sh}$ is $<\Delta\hat{H}R>_{Sh} = <\Delta HR>_{Sh}/|<\Delta HR>_A| = 10.13/15.51 = 0.653$. For PEP, we can read that $<\Delta PEP>_A = -11.29$ and $<\Delta PEP>_{Sh} = -5.51$; consequently, the corresponding normalized values are $<\Delta\hat{P}\hat{E}P>_A = -1$ and $<\Delta\hat{P}\hat{E}P>_{Sh} = -5.51/|-11.29| = -0.488$.

3. With these normalized values, the correlation factor R can be computed for all the couples of cardiovascular-emotional observables.

   For example, for the two couples (A,Sh) and (HR,PEP), the correlation factor can be computed as $R = <\Delta\hat{H}R>_A + <\Delta\hat{P}\hat{E}P>_A + <\Delta\hat{H}R>_{Sh} - <\Delta\hat{P}\hat{E}P>_{Sh} = 1.141$. The 12 values of R for this experiment are reported in the table below (Table D.3):

Table D.3 Values of R for all couples of cardiovascular/emotional observables computed from the data of (Kassam and Mendes 2013)

	A, Sh	Sh, A
HR, PEP	3.14	0.16
HR, CO	0.28	3.03
PEP, CO	−0.86	1.88
CO, PEP	0.86	−0.62
HR, CO	0.28	3.03
PEP, HR	0.86	0.16

The other parts of this meta-analysis, like those drawn from the paper by Sinha and team (Sinha et al. 1992) which involves four emotions (joy, sadness, fear and anger) and four cardiovascular variables (systolic and diastolic pressures, heart rate, stroke volume), will not be reproduced here. The complete meta-analysis is reported in the reference (Uzan 2014b).

Appendix 5: A Quantum-like Model of Bipolar Disorder

|ΦM > and |ΦD > are the two vectors of the Hilbert space H of psychosomatic states representing, respectively, the two (ideal) states of "mania" and "depression". The psychosomatic state of a bipolar patient can be approximately written as a linear combination of only these two states:

$$|\Xi >= c_D | \Phi_D > + c_M | \Phi_M >.$$

Since these two states are attractors for the patient, which has tendency to stay in one of them without external help, one can assume that they are eigenstates of the self-adjoint operator **Ω**, defined in Sect. 4, from which the unitary operator governing the rhythmic evolution of psychosomatic states is computed:

$$\Omega | \Phi_D >= \omega_D | \Phi_D > \text{and}\, \Omega | \Phi_M >= \omega_M | \Phi_M >,$$

with $\omega_D = 2\pi f_D$ and $\omega_M = 2\pi f_M$, f_D and f_M being the frequencies associated to these two states.

The psychosomatic state of a bipolar patient is thus modeled as a two-levels state governed by a unitary evolution and submitted to a perturbation. The unperturbed, rhythmic evolution is governed by the unitary operator:

$$\mathbf{R}(t + \Delta t, t) = \exp[-i\,\Omega\,\Delta t],$$

while the perturbation is represented by a hermitian operator W. Using the well-known techniques of standard quantum theory (Cohen Tannoudji et al. 1973, Chap. 4.C), it is possible to estimate, in the general case where the perturbation W is the sum of a constant operator W1 and a harmonic operator W2, the period T of alternation as follows.

P. Uzan, *Mind-Body Entanglement*, SpringerBriefs in Philosophy,
https://doi.org/10.1007/978-3-030-90173-8

Let t_0 be the characteristic period of the oscillations (roughly one year) corresponding to the evolution $R' = R + W1$ and Δt the period of the harmonic perturbation W2, which is supposed very small compared to t_0. Let us compute the probability p_N (t) for the individual to be found at an initial time t = 0 in one of these two states, for example $|\Phi_D>$, and in this same state at the later times defined by $t = n\ \Delta t$, with n ranging from 1 to N. Since the "spontaneous" oscillations with period t_0 (corresponding to the action of W1 alone) are governed by the action of the evolution operator R′, a straightforward calculation of p_N (t) leads to the following result:

$$p_N(t) = |\langle \Phi_D \,|\, R'^N \,|\, \Phi_D \rangle|^2 = \cos^{2N} \omega t, \text{with } \omega = 2\pi / t_0.$$

Expanding now p_N (t) in series and keeping only the first-order term (which is justified by $\Delta t << t_0$) leads us to write p_N (t) as:

$$p_N(t) \approx \exp\left[-4\pi^2 N(\Delta t / t_0)^2\right] = \exp\left[-4\pi^2 (\Delta t / t_0^2) t\right],$$

From this expression the mean shifting value T defined above can be computed as: $T = t_0^2/4\pi^2\ \Delta t$.

Bibliography

Aerts, D., Sozzo S., Gabora L. and Veloz T. Quantum structure in cognition: Fundamentals and Applications. *ICQNM 2011: The Fifth International Conference on Quantum, Nano and Micro Technologies,* 2011.

Alcock, J. *Parapsychologie: science ou magie.* Paris: Flammarion, 1989.

Alexander F. *Psychosomatic medicine: Its principles and applications.* New York: Norton, 1950.

Alexander F., French T. and Polock G.H. Psychosomatic specificity, vol.: Experimental studies and results. Chicago: University of Chicago Press, 1968.

American Psychiatric Association: What is Schizophrenia? https://www.psychiatry.org/patients-families/schizophrenia/what-is-schizophrenia, 2010.

Anderson S.A. The Practice of Shiatsu. Saint Louis, Missouri: Mosby Elsevier, 2008.

Ancelin-Schutzenberger A. *The Ancestor Syndrome: Transgenerational Psychotherapy and the Hidden Links in the Family Tree*. Routledge, 1998.

Aristole. *Physics* (350 BCE) *Translated by Reeve, C. D. C. Cambridge, MA: Hackett Publishing Company, 2018.*

Atmanspacher H. Mind and matter as asymptotically disjoint, inequivalent representations with broken time-reversal symmetry. *BioSystems* 68, p. 19-30, 2003.

Atmanspacher H. and Scheingraber H. (editors). *Information Dynamics*. Plenum Press, New York, 1991.

Atmanspacher H., Römer H., and Walach H. Weak quantum theory: Complementarity and entanglement in physics and beyond. *Foundation of Physics* 32, 379–406, 2002.

Atmanspacher H., Filk T. and Römer H. Théorie quantique faible: cadre formel et applications, *in M. Bitbol (ed.), Théorie quantique et sciences humaines,* CNRS Editions, 2009.

Armstrong D.M. *A Materialist Theory of the Mind*, London, Routledge, 1968.

Aspect A., Dalibard J. and Roger G. Experimental Test of Bell's Inequalities Using Time-Varying Analyzers, *Phys. Rev. Lett.*, 49 (25): 1804–7, 1982.

Augustine of Hippo *Confessions* (397-400 AD), Trans. Vernon J. Bourke. Washington: Catholic University of America Press, 1966.

Baillarger J. Notes on a type of insanity in which attacks are characterised by two regular periods, one of depression, the other of excitation. French Academy of Medicine, 30 January 1854. Acad. Med. Bull. (Paris);19: 340–352, 1854.

Balaban R.S. Metabolic homeostasis of the heart. *Symposium on Mitochondrial Physiology and Medicine,* May 29, 2012.

Baptista J. Derakhani M. and Trssoldi P. Explicit Anomalous Cognition. A Review of the Best Evidence in Ganzfeld, Forced–Choice, Remote Viewing and Dream Studies. In Etzel C.,

P. Uzan, *Mind-Body Entanglement*, SpringerBriefs in Philosophy,
https://doi.org/10.1007/978-3-030-90173-8

Plmer J. and Marcusson-Clavertz D. eds. *Parapsychology. A Handbook for the 21th Century*. McFarland, 2015.

Barber T.X. Hypnosis, suggestion and psychosomatic phenomena. *American Journal of Clinical Hypnosis*. 21: 12–27, 1978.

Bard, P. Emotion: I. The Neuro-humoral Basis of Emotional Reactions. In C. Murchison (Ed.), *A handbook of general experimental psychology* (pp. 264–311). Clark University Press, 1934.

Barry J. General and comparative study of the psychokinetic effect of a fungus culture. *J. of Parapsychology* 32: 327-243, 1968.

Bateson G. *Mind and Nature: A Necessary Unity*, 1979. French translation *La nature et la pensée*, Seuil, Paris, 1984.

Bateson G. La nature et la pensée. *Esprit et nature, une unité nécessaire*. Paris: Seuil, 1984.

Baumgartner M. Interventionism and Epiphenomenalism, *Canadian Journal of Philosophy* 40: 359-383, 2010.

Beebee H. Hume and the Problem of Causation, in P. Russell (ed.), *The Oxford Handbook of Hume,* New York: Oxford University Press, 2016.

Bell J.S. On the Einstein Podolsky Rosen paradox. Physics Vol. 1, No. 3, pp. 195-290, 1964.

Bengston W. and Krinsley D. The Effect of the "Laying On of Hands" on Transplanted Breast Cancer in Mice. *Journal of Scientific Exploration* 14(3): 353–364, 2000.

Benson H. *The Relaxation Response*. New York: Morrow 1975.

Benson H., Dusek J.A., Sherwood J.B., *et al.* Study of Therapeutic Effects of Intercessory Prayer (STEP) in cardiac bypass patients: a multicenter randomized trial of uncertainty and certainty of receiving intercessory prayer. American Heart Journal 151: 934-942, 2006.

Benor D.J. *Spiritual Healing, Scientific Validation of a Healing Revolution*. Southefield, Michigan: Vision Publications, 2001.

Bergson H. *Durée et Simultanéité* (1922). Re-ed. Presses Universitaires de France, 1998.

Bernard C. Leçons sur la chaleur animale, Paris: J.-B. Baillière, 1876.

Berg N.J. The Creator in creation. In Branover H., Attia I.C. eds., *Science in the Light of the Torah*. NJ: Jason Aroson, 1994.

von Bertalanffy L. *General Systems Theory* (1968) French translation Jean-Benoît Chabrol. *Théorie Générale des Systèmes*. Paris: Dunod 2012.

Beugnet-Lambert C. *Chronopsychologie: rythmes et activités humaines*. Presses Universitaires de Lille, 1988.

Bitbol M. *Mécanique quantique. Une introduction philosophique*, Paris, Flammarion, 1996.

Black P. *Physiological Correlates of Emotion*. Academic: New York and London, 1970.

Bleich J.D. Maimonides on the distinction between science and pseudoscience. In Rosner F., Kottek S.S eds. Moses Maimonides: Physician, Scientist and Philosophers. Northvale, NJ: Jason Aronson, 105-115, 1993.

Bohm D. *Wholeness and the Implicate Order*. Routledge and Kegan Paul, 1980.

Bohm D. and Hiley B. *The Undivided Universe*. London and New York: Routledge, 1993.

Bougeois ML *Manie et dépression. Comprendre et soigner le trouble bipolaire*. Paris: Odile Jacob, 2011.

Bourgeois M.L., Gay C., Chantal H. and Masson M. *Les troubles bipolaires*. Coll Psychiatrie. Paris: Lavoisier, 2014.

Braud W.G. Lability and inertia in conformance behaviour. Journal of American Society for Psychical Research 74: 297-318, 1990.

Broom B. Meaning-Full Disease: How Personal Experience and Meanings Cause and Maintain Physical Illness. First published by Karnac Books, 2007. New York: Routledge 2018.

Bub, J. Why the Tsirelson Bound? In: Ben-Menahem, Y. and Hemmo, M., Eds., Probability in Physics, The Frontiers Collection, Springer, Berlin, Heidelberg, 167-185, 2012. https://doi.org/10.1007/978-3-642-21329-8_11

Busemeyer J.R. and Bruza P.D. *Quantum Models of Cognition and Decision*. Cambridge University Press, 2012.

Butterfield J. Laws, causation and dynamics at different levels. Interface Focus 2(1): 101–114, 2012.

Byrd R.C. Positive therapeutic effects of intercessory prayer in a coronary care population. *Southern Medical J.* 81(7): 826-829, 1988.
Canguilhem G. Le Normal et le pathologique, Paris, Presses Universitaires de France, 1966.
Cannon W. B. The James-Lange theory of emotions: A critical examination and an alternative theory. *American Journal of Psychology, 39*, 106-124, 1927.
Cannon W.B. The James-Lange theory of emotions: A critical examination and an alternative theory. Special Centennial Issue *Am J Psychol* 1987;100(3/4):567. doi:https://doi.org/10.2307/1422695
Cerutti E. *Mystic With the Healing Hands: The Life Story of Olga Worall.* Ney York: Harper, 1975.
Chalmers, D. *The Conscious Mind.* Oxford: Oxford University Press, 1996.
Chalmers, D. The Hard Problem of Consciousness, in M. Velmans and S. Schneider *The Blackwell Companion to Consciousness,* Blackwell Publishing, pp. 225-235, 2007.
Chiarioni G. and Whitehead W.E. The role of biofeedback in the treatment of gastrointestinal disorders. *Nature Clinical Practice Gastroenterology & Hepatology* 5: 371–382, 2008.
Chiozza A. *Corpo, affetto e linguaggio.* Turin: Loescher, 1976.
Cho Z.H., Chung S.C, Jones J.P, Park J.B., Lee H.J., Wong E.K. and Min B.I. New findings of the correlations between acupoints and corresponding brain cortices using functional MRI. *Proceedings of the National Academy of Sciences*, 95: 2670-2673, 1998.
Chopra A. and Doiphode V.V. Ayurvedic medicine. Core concept, therapeutic principles, and current relevance. Medical Clinics of North America 86(1):75-89, vii, 2002.
Churchland P. M. The Engine of Reason, the Seat of the Soul. Cambridge, MA: MIT Press, 1996.
Clark A. and Chalmers D.J. The extended mind. *Analysis.* 58(1): 7–19, 1998.
Clauser J. F., Horne M. A., Shimony A. and Holt R. A. Proposed Experiment to Test Local Hidden-Variable Theories. *Phys. Rev. Lett.* (23): 880-890, 1969.
Cohen-Tannoudji C, Diu B, Laloë F. Mécanique Quantique. Hermann, 1973.
Collipp P.J. The Efficacy of Prayer: A Triple blind Study. *Medical Times* 97(5): 201-204, 1969.
Cooper M.J. and Aygen M.M. A relaxation technique in the management of hypercholesterolemia. *J Human Stress* 5(4):24-7; 1979.
Crane, T. *Elements of Mind: An Introduction to the Philosophy of Mind*, Oxford: Oxford University Press, 2001.
Crick, F. *Astonishing Hypothesis: The Scientific Search for the Soul.* New York, Scribner, 1994.
Damasio A. *The feeling of what happens: Body et Emotions in the making of Consciousness.* New York: Harcourt Brace, 1999. French trans. *Le sentiment même de soi. Corps, émotions, conscience.* Paris: Odile Jacob, 1999.
Damasio A. *Descartes' Error: Emotion, Reason and the Human Brain.* New York, Avon Books, 1994.
Darwin C. *The Expression of the Emotions in Man and Animals. London:* John Murray, 1872,
Davidson, D. Mental Events, in *Essays on Actions and Events*, Oxford: Clarendon Press, 1980.
Davidson R.J., Kabat-Zinn J, et al. (7 authors) Alterations in Brain and Immune function produced by mindfulness meditation. *Psychosomatic Medicine* 65: 564-570, 2003.
Dawkins R. Is Science a Religion? the *Humanist*, January/February 1997.
Dehaene S. and Changeux J. P. Neural mechanisms for access to consciousness, *in* Michael Gazzaniga et al., *The Cognitive Neuroscience III.* Cambridge: MIT Press, 2004.
Dennett, D. *Consciousness Explained.* Boston: Little, Brown, 1991.
Descartes, R. *Treatise of Man.* French translation: *Traité de l'Homme*, 1664. website Gallica: http://gallica.bnf.fr/ark:/12148/bpt6k574850/f143.
Descartes R. Lettre au Marquis de Newcastle du 23 novembre, in *Oeuvres et lettres*, Gallimard, coll. "Bibliothèque de la Pléiade", p. 1254-1257, 1953.
Descartes R. The Philosophical Writings of Descartes, trad. John Cottingham, Robert Stoothoff, Dugald Murdoch, Cambridge University Press, 1984.
Descartes R. *Traité des passions,* 1649. Reedition Paris, Vrin 1994.
Descartes R. *Principia philosophiae,* 1644. Reedition: *Principes de philosophie* Paris, Vrin 2000.

Dossey, Larry *Healing Words: The Power of Prayer and the Practice of Medicine*. US: HarperSanFrancisco Publisher, 1993.

Dossey L. Healing Research: What We Know and Don't Know. Explore Vol. 4 (6), p. 341-351, 2008.

Dossey L. Nonlocal Mind: a (Fairly) Brief History of the Term. *Explore* vol. 11 no. 2, 2015.

Duggan M. and Tressoldi P. Predictive physiological anticipatory activity preceding seemingly unpredictable stimuli: An update of Mossbridge *et al*'s meta-analysis 2018. F1000Research 7:407, 2018.

Eccles, J. *Evolution du cerveau et création de la conscience*, Champs Flammarion, Paris, 1994.

Eells E. *Probabilistic Causality*, Cambridge: Cambridge University Press, 1991.

Einstein A., Podolski, B. and Rosen N. Can Quantum-Mechanical Description of Physical Reality Be Considered Complete? *Physical Review* 47, 777-780, 1935.

Ellsberg D. Risk, Ambiguity, and the Savage Axioms. Quarterly Journal of Economics. 75 (4): 643–669, 1961.

Engel G. L. The need for a new medical model: A challenge for biomedicine. *Science* 198:129-96, 1977.

Ernst E., Rand J.I. and Stevinson C. Complementary therapies for depression: an overview. *Archives of General Psychiatry* 55: 1026-1032, 1998.

Esfeld, M. *La philosophie de l'esprit*. Paris: Armand Colin, 2005.

Espagnat (d') B., *Le réel Voilé*. Paris: Fayard, 1994.

Fechner G. T. *Elemente der Psychophysik*. Leipzig: Breitkopf & Hartel, 1860.

Fechner G.T. Zend-Avesta. Leipzig: Voss., 1851

Feigl, H. The 'Mental' and the 'Physical', in Feigl H., Scriven, M. and Maxwell G., *Minnesota Studies in the Philosophy of Science: Concepts, Theories, and the Mind-Body Problem*, Minneapolis: Minnesota University Press, 370–409, 1958.

Feldenkrais M. *Awareness Through Movement*. New-York: Harper and Row, 1967.

Février P., Barreau H. and Lochack G. (editors) *J.L. Destouches Physicien et Philosophe*, CNRS éditions 1994.

Fiammetti R. *Le langage émotionnel du corps*. Dervy-Livres, 2004.

Friedman B.H. Feelings and the body: The Jamesian perspective on autonomic specificity of emotion. *Biological Psychology*. 84:383–93, 2010.

Finger F.W. Circadian rhythms: implications for psychology. N.Z. Psychol 11(1):1–12, 1982.

Fleckenstein J., Krüger P. and Ittner K. Effects of single-point acupuncture (HT7) in the prevention of test anxiety: Results of a RCT. PLoS One, Aug 30, 13(8), 2018.

Foster J. *The Immaterial Self*. London Routledge, 1991.

Frege G. *Über Sinn und Bedeutung*. In: *Zeitschrift für Philosophie und philosophische Kritik*. Band 100, 1892, S. 25–50, 1892. French translation Claude Imbert, Sens et dénotation , in *Écrits logiques et philosophiques*, pp. 102-126, Paris: Seuil, 1971.

Freud S. Fragment d'une analyse d'hystérie (Dora). *In* Freud S. *Cinq psychanalyses* (trad. M. Bonaparte, R. M. Lœwenstein): 1-91. Paris: Presses Universitaires de France, 1905a.

Freud S. *Drei Abhandlungen zur Sexualtheorie* (1905b). French translation Reverchon-Jouve B. *Trois essais sur la théorie de la sexualité*, Paris: Gallimard, 1923

Freud, S. *Studies on Hysteria* (1895). Standard Edition, vol. 2, Hogarth Press, 1955.

Freud S. Treibe und Triebschicksale (1915). French translation Laplanche J. and Pontalis J.B. *Métapsychologie*. Paris: Gallimard, 1965.

Friedman M. and Rosenman R. *Type A Behaviour and Your Heart*. New York: Knopf, 1974.

Fuente-Fernandez R, Lidstone S, Stoessl AJ Placebo effect and dopamine release. J Neural Transm Supplementa 70:415–418, 2006.

Fusswerk-Fursay, J. La notion de croyance et son importance en psychiatrie. *Revue d'Histoire de la Médecine Hébraïque*, 124: 17-20, 1978.

Galton F. Statistical Studies into Efficacy of Prayer. *Fortnightly Review* 12: 125-135, 1872.

Gasking, Douglas Causation and Recipes, *Mind* 64, p. 479-487, 1955.

Gejman P.V., Sanders A.R., Duan J. 2010. The role of genetics in the etiology of schizophrenia. *Psychiatr Clin North Am.* 33(1):35-66.

Geringer J.M., Madsen K.M. and Gregory D. A fifteen-year history of the Continuous Response Digital Interface: Issues relating to validity and reliability, *Bul. Council for Research in Music Education,* 2004.

Gisin N. Bell's inequality holds for non-product states. *Physics Letters A* 154, 201-202, 1991

Gleason A. Measures on the closed subspaces of a Hilbert space. Journal of Mathematics and Mechanics 6, 885-893, 1957.

Gomes G. The timing of conscious experience: a critical review and reinterpretation of Libet's research, Consciousness & Cognition 7: 559-595, 1998.

Goodkin K. and Visser A.P., eds. *Psychoneuroimmunology: Stress, Mental Disorders And Health,* Ney York: American Psychiatric Press, 2000.

Grinker, RR Sr. A struggle for eclecticism. *Am J Psychiatry* 7 (121), 1964.

Groddeck G. *Das Buch Vom Es* (1923), Translated from German to French by Lily Jumel: *Le livre du ça.* Paris: Gallimard 1973.

Groddeck, G. The Meaning of Illness. The Hogarth Press and the Institute of Psychoanalysis, London, 1928.

Hamer, R. G. (2005) The Five Biological Laws of the New Medicine. *First International Congress on Complementary and Alternative Medical Cancer Treatment.* May 14/15, Madrid, Spain.

Hammer B.U., Colbert A.P., Brown K.A., Ilioi E.C. Neurofeedback for insomnia: a pilot study of Z-score SMR and individualized protocols, *Appl Psychophysiol Biofeedback* 36(4):251-64, December 2011.

Harris W., Gowda M., Kolb J.W., Strychacz C.P., Vacek J.L., Jones P.G., Forker A., O'Keefe J.H. and McCallister B.D. A randomized, controlled trial of the effects of remote, intercessory prayer on outcomes in patients admitted to the coronary care unit. *Archives of Internal Medicine* 159, 19, pp. 2273-2278, 1999.

Horowitz M.J. Image Formation and Psychotherapy. New York: Aronson, 1983.

Haynes S., Fenleb M. and Levine S. The relationship of psychosocial factors to coronary heart disease in the Framingham Study. *American Journal of Epidemiology* 107: 384-389, 1978.

Healy D. *Psychiatric drugs explained.* London, Elsevier 2008.

Hempel C.G. Studies in the Logic of Explanation, in Carl Gustav Hempel, Aspects of Scientific Explanation, Ney York: The Free Press, 1965.

Hilgard J. Anniversary Reactions in Parents Precipitated by Children, *Psychiatry,* 16: 73-80, 1953.

Hofstadter D. and Dennett D. *Vues de l'esprit.* Intereditions, 1987.

Hofmann, S. G. and Gómez, A. F. Mindfulness-Based Interventions for Anxiety and. Depression. *Psychiatric Clinics of North America,* 40(4), 739–749, 2017.

Hughes R.I.G. *The structure and interpretation of quantum mechanics,* Harvard University Press, 1989.

Hultkrantz A. Shamanic Healing and Ritual Drama: Health and Medicine in Native North American Religious Traditions. Crossroad Publishing Co, U.S, 1997.

Hume D. *A Treatise of Human Nature* (1740), ed. L. A. Selby-Bigge, 2nd ed. Revised by P. H. Nidditch, Oxford: Clarendon Press, 1975.

Illari, Ph. and Jon, W. What is a Mechanism? Thinking About Mechanisms Across the Sciences. *European Journal for the Philosophy of Science* 2: 119-35, 2012.

INSERM Press Office 2018: The Biological Clock Sets a Different Rhythm for Each Organ. https://presse.inserm.fr/en/the-biological-clock-sets-a-different-rhythm-for-each-organ/30595/

Jackson, F. Mental Causation, *Mind,* 105: 377–413, 1996.

Jackson D.D. *Therapy, Communication and Change* (Human communication, volume 2), Palo Alto, CA: Science & Behaviour Books, 1968 Palo Alto, CA: *Science & Behaviour,* 1968.

Janov A. *The Biology of love.* New York: Prometheus Books, 2000.

Janssen T. *La solution intérieure.* Paris: Fayard, 2006.

Johnstone E.C., Frith C.D., Crow T.J., Owens D.C. The Northwick Park "functional psychosis study: diagnosis and treatment response. *Lancet,* Jul 16;2 (8603):119-25, 1988.

Jonas H. *Le phénomène de la vie.* French translation by Danielle Lorries Paris: De Boeck, 1982.

Jonas H. 1986. Parallelism and Complementarity: The Psycho-Physical Problem in Spinoza and in the Succession of Niels Bohr. In: Grene M., Nails D. eds: *Spinoza and the Sciences*. Boston Studies in the Philosophy of Science, vol 91. Springer, Dordrecht

Jonas W.B. and Crawford C.C. *Healing, Intention and Energy Medicine.* New York: Churchill Livingstone, 2003.

Jung C.G. Preface of *I Ching or the Book of Changes*, traduction de Cary F. Baynes, New York Londres, 1951.

Jung, C.G. *Mysterium Conjunctionis, in* The Collected Works of C.G. Jung (Volume 14), London: Routledge, 1963.

Jung, C.G. *Mind and his symbols*. Ferguson Publishing Company. French Trans. *L'homme et ses symboles*. Paris: Robert Laffont, 1964.

Kabat-Zinn, J. Wherever you go. There you are: mindfulness meditation in everyday life. New-York, United State: Hachette Book, 1994.

Kandel E. *In Search of Memory.* Norton, 2006. French translation, Paris: Odile Jacob 2007.

Kassam K.S. and Mendes W. B. The Effects of Measuring Emotion: Physiological Reactions to Emotional Situations Depend on whether Someone Is Asking. *Plos One*, June 05, 2013.

Khayat D. *De larmes et de sang*. Odile Jacob, 2013.

Kim, J. On the Psycho-Physical Identity Theory, American Philosophical Quarterly (3): 227-235, 1966.

Kim, J. *Supervenience and Mind: Selected Philosophical Essays*. Cambridge University Press 280, 1993.

Kim, J. *Physicalism, or something near enough.* Princeton University Press, 2005.

King S. Kahuna Healing: Holistic Health and Healing Practices of Polynesia. Quest Books, U.S, 1996.

Kistler, M. Reducing Causality to Transmission, *Erkenntnis* (48): 1-24, 1998.

Kistler, M. La causalité dans la philosophie contemporaine.Intellectica 38:139-185, 2004.

Kistler, M. The Interventionist Account of Causation and Non-causal Association Laws. *Erkenntnis* 78: 65–84, 2013.

Koch, C. *A la recherche de la conscience.* Paris: Odile Jacob (2006) (French) trad. from *The Quest for Consciousness* 2004.

Kraft D. Portrait of a Psychic Healer. New York: G.P. Ptnam's Sons, 1981.

Kranitz L. and Lehrer P. Biofeedback Applications in the Treatment of Cardiovascular Diseases. *Cardiol Rev.*12(3):177-81, 2004.

Kraepelin E. *Manic Depressive Insanity and Paranoia*. Edinburgh: E. & S. Edinburgh: Livingstone; 1921

Kreibig S.D., Wilhelm F.H., Roth W. and Gross J.J. Cardiovascular, electrodermal, and respiratory response patterns to fear- and sadness-induced films. *Psychophysiology* 44: 787-806, 2007.

Krucoff M.W., Crater S.W., Green C.L., Maas A.C., Seskevich J.E., Lane J.D., Loeffler K.A., Morris K., Bashore T.M. and Koenig H.G.. Integrative noetic therapies as adjuncts to percutaneous intervention during unstable coronary syndromes: Monitoring and Actualization of Noetic Training (MANTRA) feasibility pilot, *American Heart Journal* 142, 5, pp. 760-767, 2001.

Kubik, A. and Biedroń, A. Neurofeedback therapy in patients with acute and chronic pain syndromes – literature review and own experience. *Przegl Lek* 70(7): 440-2, 2013.

Kranitz L. and Lehrer P. Biofeedback applications in the treatment of cardiovascular diseases. Cardiol Rev. 12(3):177-81, 2004.

Kripke S. 'Naming and Necessity'. In Davidson, Donald and Harman, Gilbert, eds., *Semantics of Natural Language*. Dordrecht: Reidel, 1972.

Lambert J.F. Les miracles entre Science et Foi, Expliquer ou Comprendre. *Connaître* no. 51, 2018.

Landau L. J. Experimental tests of general quantum theories. *Lett. Mathematical Physics* 4: 33-40, 1987.

Landau L.J. *Quantum theory: Mathematics and reality*, Chapter 7. Talk Given at the King's College History of Mathematics Summer School, King's College London. Manuscript available at www.mth.kcl.ac.uk/events/summer_schools/summer_school2001/qschool.html, 2001.

Lang P.J. A bio-informational theory of emotional imagery. *Psychophysiology* 16:495-512, 1979.
Laplantine F. *Anthropologie de la maladie*, Paris: Payot, 1993.
Laureys S. La Méditation c'est bon pour le cerveau, Odile Jacob, 2019.
Lawrence B. *The Practice of the Presence of God*, 1692. Translated by Edmonson R.J., edited by Helms H.M, Orleans, Mass: Paraclete Press, 1985.
Lazarus R. *Emotion and adaptation*. New York: Oxford University Press, 1991.
Leibovici L. Effects of remote, retroactive intercessory prayer on outcomes in patients with bloodstream infection: randomized control trial. *British Medical Journal* 323 (7327): 1450-1, 2001.
Lestienne R. Les *fils du temps*: *causalité, entropie, devenir.* Presses du CNRS, 1990.
Lestienne R. *Miroirs et tiroirs de l'âme*. CNRS Editions, 2008.
Lestienne R. *Dialogues sur l'émergence*, Le Pommier, 2012.
Levin J.S. How Prayer Heals: A Theoretical Model. *Alternative Therapies in Health and Medicine* (1): 66-73, 1996.
Levin J.S. and Vanderpool H.Y. Is religion therapeutically significant for hypertension? Soc. Sci. Med. 29: 69-78, 1989.
Levine, J. Materialism and Qualia: The Explanatory Gap. *Pacific Philosophical Quaterly* 64, pp. 354-361, 1983.
Libet B. Unconscious cerebral initiative and the role of conscious will in voluntary action. *Behavioural and Brain Research* 8: 529-566, 1985.
Leopold D.A. and Logothetis N.K. Activity changes in early visual cortex reflects monkeys' percepts during binocular rivalry. *Nature 379:* 549-553, 1996.
Leshan L. and Margeneau H. An approach to a science of psychical research. *J. of the society of Psychical Research* 50: 273-283, 1980.
Leventhal H. and Scherer K. The Relationship of Emotion to Cognition: A Functional Approach to a Semantic Controversy. *Cognition and Emotion* 1(1):3-28, 1987.
Lewis D. Causation, in *Philosophical Papers*, Vol. II: 159-213, New York: Oxford University Press, 1986.
List C. and Menzies P. Non-reductive physicalism and the limits of the exclusion principle. *J. Phil.* 106, 475–502, 2009.
Littman A.B. Review of psychosomatic aspects of cardiovascular disease. Psychotherapy and Psychosomatics 60 (3-4): 148-167, 1993.
Lucadou W. Verhexung – Erfahrungen einer parapsychologischen Beratungsstelle. [Bewitchment – Experiences in a parapsychological counselling office]. In W. Bruchhausen (Ed.), *Hexerei und Krankheit [Witchery and Disease]*, 2003 (pp. 195 – 218).
Lucadou W., Römer, H., and Walach H.: Synchronistic phenomena as entanglement correlations in generalized quantum theory, *Journal of Consciousness Studies,* Band 14, 2007.
Lucadou W. Homoeopathy, Macroscopic Entanglement, and the Model of Pseudo-Machines. Intervention at the International meeting on 'Homeopathy Mechanism, Methods and Measurement', Paris, September 2019.
Luo H.C., Jia Y.K. et al. Electroacupuncture vs Ametriptyline in the treatment of depressive states. Journal of Traditional Chinese Medicine 5: 3-8, 1985.
Lunel P. *Les guérisons miraculeuses. Enquête sur un phénomène inexpliqué*, J'ai lu, 2004.
Lutz A., Greischar L.L., Rawlings N.B., Ricard M. and Davidson R.J Long-term meditators self-induce high-amplitude gamma synchrony during mental practice. PNAS 101 (46), 2004.
Lyall LM, Wyse CA, Graham N, et al. Association of disrupted circadian rhythmicity with mood disorders, subjective wellbeing, and cognitive function: a cross-sectional study of 91 105 participants from the UK Biobank. *The lancet Psychiatry* 5(6):507-514, 2018.
Lycan W.G. Giving Dualism its Due. Australasian Journal of Philosophy 87 (4): 551-563, 2009.
Lytle C. *An Overview of Acupuncture.* Washington, DC: United States Department of Health and Human Services, 1993.
Mach E. *Die Analyse der Empfindungen*, Iena, G. Fischer. 1886 Traduction F. Eggens et J.-M. Monnoyer: *L'Analyse des sensations*, Nîmes, J. Chambon, 1996.

Maclean, P. D. Psychosomatic disease and the "visceral brain"; recent developments bearing on the Papez theory of emotion. *Psychosomatic Medicine,* 11: 338–353, *1949.*

Maharishi Mahesh Yogi. *Science of Being and Art of Living: Transcendental Meditation.* New York, Signet 1963.

Malanga C.J. The Comprehensive Pharmacology Reference, 2007

Maranon G. Contribution à l'étude de l'action émotive de l'adrénaline. Review Francaise Endocrinology 2:301-325, 1924.

Marty P., de M'Uzan, M. and David, C. *L'investigation psychosomatique.* Paris: Presses Universitaires de France, 1963.

Mc Dougall J. Théâtre du corps. Paris: Gallimard, 1986.

Mc Laughlin, B. and Bennett, K. Supervenience, *Stanford Encyclopedia of Philosophy.* http:// plato.stanford.edu, 2011.

Mc Ginnis M. *Maidens, Matrons, and Magicians: Women and Personal Ritual Power in Late Antique Egypt.* Thesis presented for the degree of Master of Arts, University of Louisville, Kentucky, 2012.

Mc Manaway B. and Turcan J. *The Energy That Can Restore Health.* Wellingborough, England: Thorsons, 1983.

Mc Taggart J. The Unreality of Time. *Mind: A Quarterly Review of Psychology and Philosophy* 17, 456-47, 1908.

Mills, E. "Interactionism and Overdetermination", *American Philosophical Quarterly,* 33: 105–17, 1996.

Moravec C.S. and McKee M.G. Biofeedback in the treatment of heart disease (Cleveland Clinic) *Journal of Medicine*, 78 Suppl 1:S20-3, August 2011.

Mossbridge J, Tressoldi P and Utts J. Predictive physiological anticipation preceding seemingly unpredictable stimuli: a meta-analysis. *Frontiers in Psychology* 3:390, 2012.

Meier, C.A. (ed.) Atom and Archetype. The Pauli/Jung Letters, 1932–1958. Princeton University Press, Princeton, 2001.

von Meyenn K. (ed.) Wolfgang Pauli. Wissenschaftlicher Briefwechsel, Band III: 1940–1949. Springer, Berlin, 1993.

Minsky M. *La Société de l'Esprit*, Paris: InterEditions, 1988a.

Mullally, W.J., Hall, K. & Goldstein, R. Efficacy of biofeedback in the treatment of migraine and tension type headaches. *Pain Physician* 12(6):1005-11), Nov-Dec 2009.

Mohr H.M. Orientation-specific aftereffects to mentally generated lines. *Perception* 40:272–290, 2011.

Mattuck, R. Une théorie quantique de l'interaction entre la conscience et la matière. In *Science et Conscience, les deux lectures de l'univers,* Editions Stock, Paris, pp. 77-96, 1980.

Minsky M. *La Société de l'Esprit, Paris: InterEditions,* 1988b.

Nagel, T. What Is It Like to Be a Bat? *Philosophical Review* (83):435-450, 1974.

Nash C.B. Psychokinetic control of bacterial growth. *J. of the society of Psychical Research.* 51: 217-221, 1982

Nash C.B. Test of psychokinetic control for bacterial mutation. *J. of the society of Psychical Research* 78(2): 145-152, 1984.

von Neumann J. *Mathematical Foundations of Quantum Mechanics.* Princeton University Press., 1955.

Noë A. and Thompson E. Are there Neural Correlates of Consciousness? *Journal of Consciousness Studies*, 11, No. 1, pp. 3–28, 2004.

Nummenmaa L., Glerean E., Hari R. and Hietanen J.K. Bodily maps of emotions. *PNAS*, 111 (2) 646-651, January 14, 2013.

Olwen G.O. Discovering a seasonal pattern in bipolar disorder. Med News Today (MNT), 2007. http://www.medicalnewstoday.com/articles/866(20.php

Papineau, D. *Philosophical Naturalism*, Oxford: Blackwell, 1993.

Patouillard R. La schizophrénie: cent ans encore ? *La revue lacanienne, 2* (10): 105-112, 2011.

Pauli W. letter no 1391 to Rosenfield, 1er Avril 1952, *in* von Meyenn: *Wolfgang Pauli.*

Wissenschaftlicher Briefwechsel, Band IV, p. 593. Berlin: Springer, 1996.

Pauls C.A. and Stemmler G. Repressive and Defensive Coping during Fear and Anger. *Emotion*, Vol. 3(3): 284-302, 2003.

Pawlowski M.P., Paterek T., Kaszlikowski, D., Scarani, V., Winter, A. and Zukowski, M. Information Causality as a Physical Principle. Nature, 461, 1101-1104, 2010.

Petrovic. P, Kalso. E, Petersson. KM, Ingvar M. Placebo and opioid analgesia—imaging a shared neuronal network. *Science* 295: 1737-1740, 2002.

Petter M., McGrath P. J., Chambers C. T. and Dick B. D. The effects of mindful attention and state mindfulness on acute experimental pain among adolescents. *Journal of Pediatric Psychology* 39(5), 521-531, 2014.

Peters D. *Understanding the Placebo Effect in Complementary Medicine; Theory, Practice and Research.* Churchill Livingstone, 2001.

Peters A. and Kara D.A. The neuronal composition of area 17 of rat visual cortex. IV. The organisation of pyramidal cells. *Journal of Comparative Biology* 260: 573-590, 1987.

Pfeffer M., Korf H-W and Wicht H. Synchronizing effects of melatonin on diurnal and circadian rhythms. *General and Comparative Endocrinology* 258, 2017.

Philippidis C., Dewdney B. J. and Hiley B. Quantum interference and the quantum potential, *Il Nuovo Cimento, B*, vol. 52, no. 1, p. 15-28, 1979.

Platz T., Junkert B. and Tress W. *Is there a typical conflict in gastric complaints? Psychotherapy and Psychosomatics* 59: 165-172, 1993.

Polkinghorne J. *Science and Providence: God's Interactions with the World.* Boston, Mass: New Science Library, 1989.

Popper K., Peter Arhem, Ingemar B. Lindahl, A Discussion of the Mind-Brain Problem, *Theoretical Medicine*, vol. 14, 1993.

Price, Huw Agency and Causal Asymmetry, *Mind* 101: 501-520, 1992.

Primas H. Synchronizität und Zuffal. Zeitschr. *Grenzgeb. Psych.* 38, p. 61-91, 1996.

Primas H. Time-entanglement between mind and matter. *Mind and Matter 1*, 81-119, 2003.

Primas H. Complementarity of Mind and Matter, in *Recasting Reality*, Atmanspacher H and Primas H. editors, 2010.

Prkachin K. M., Williams-Avery R.M., Zwaal C. and Mills D.E. Cardiovascular Changes During Induced Emotion: An Application of Lang's Theory of Emotional Imagery. Journal of Psychosomatic Research, Vol. 47(3): 255-267, 1999.

Putnam H. Psychological Predicates, in W.H. Capitan and D.D. Merrill (eds.), *Art, Mind, and Religion*, Pittsburgh: University of Pittsburgh Press, 37–48, 1967.

Radin D. *Entangled Minds: Extrasensory Experiences in a Quantum Reality*. Simon and Schuster, 2006.

Radin D. Consciousness and Our Entangled Reality. *Explore* Vol 3 (6), pp. 605-612, 2007.

Radin D., Schlitz M. and Baur CH. Distant Healing Intention Therapies: An Overview of the Scientific Evidence. *Global Advances in Health Medicine* 4 (suppl.): 67-71, 2015.

Radin D. and Nelson R. Evidence for Consciousness-Related anomalies in random Physical Systems. *Foundations of Physics*. Vol 19 (12), 1989.

Rakel D. Integrative Medicine, 2007.

Read J. Biological psychiatry's lost cause. In J. Read L.R., and Bentall R.P. (eds), *Models of madness: Psychological, social and biological approaches to schizophrenia*. New-Yok, Routledge, 2004.

Reich W. (1933) *Charakteranalyse*, English Translation: *Character Analysis*, Simon and Schuster, 1972.

Reichenbach H. The Direction of Time (1956), Berkeley: University of California Press, 1991.

Reinberg, A., and Smolensky, M. H. *Biological Rhythms* and Medicine. Cellular, Metabolic, Physiopathologic, and Pharmacologic Aspects. New York: Springer, 1983.

Rhine J.B. Extrasensory Perception. Boston: Branden, 1964.

Ribot T. *La psychologie des sentiments*. Paris: Alcan, 1896.

Richmond N. Two series of PK tests on paramecia. *J. of the society of Psychical Research* 36: 577-578, 1952.

Römer H. *Complementarity of process and substance. Mind Matter 4(1):69–89, 2006.*

Rommeluère E. *S'asseoir tout simplement: L'Art de la méditation zen*. Seuil, 2015

Rosenkranz M.A., Jackson D.C. Dalton K.M. Affective style and in vivo immune response: neurobehavioural mechanisms. Proceedings of the National Academy of Sciences (100): 1148-1152, 2003.

Roy, J.M. Naturalisme émergentiste et explication causale. *Intellectica,* 2004/2, 39: 199-227, 2004.

Russell, B. *The Analysis of Mind*, London: George Allen & Unwin, 1921.

Russell, B. *Human Knowledge: Its Scope and Limits*, London: George Allen & Unwin, 1948.

Ryu H., Jun C.D., Lee B.S. *et al.* Effect of Qigong training on proportion of lymphocyte subsets in human peripheral blood. American Journal of Chinese Medicine 23: 27-36, 1995.

Salmon W. *Scientific Explanation and the Causal Structure of the World*, Princeton: Princeton University Press, 1984.

Savage L. *The Foundations of Statistics*. New York, Wiley, 1954.

Schachter S. and Singer J. Cognitive, social and physiological determinants of emotional state. *Psychol. Rev.* 69: 379-399, 1962.

Sheldrake R. *The Presence of the Past: Morphic Resonance and the Memory of Nature.* Park Street Press, 1995.

Schneider R. H., Alexander C. N., Staggers F., Orme-Johnson, D. W., Rainforth, M., Salerno, J. W., et al. A randomized controlled trial of stress reduction in African Americans treated for hypertension for over one year. *American Journal of Hypertension*, 18, 88–98, 2005.

Schrödinger, E. Discussion of Probability Relations between Separated Systems, *Proceedings of the Cambridge Philosophical Society* (31): 555–563, (1935).

Searle J. *The Mystery of Consciousness.* Granta Books publisher, 1997. French transl. C. Tiercelin: Le mystère de la conscience, Odile Jacob, 1999.

Selye H. *The stress of life*. New York: Mc Graw-Hill, 1978.

Shapiro, L., and Sober, E. Epiphenomenalism I the Do's and the Don'ts, in P. Machamer and G. Wolters (eds.), *Thinking about Causes*, University of Pittsburgh Press, pp. 235-264, 2007.

Schmidt H. Comparison of PK action on two different random generators. *J. of Parapsychology* 38: 47-55, 1974.

Schmidt H. PK-effect on pre-recorded targets. *J. of the American Society for Psychical Research* 70: 267-291, 1976.

Schloesser R.J., Huang J., Klein P.S. and Manji H.K. Cellular plasticity cascades in the pathophysiology and treatment of bipolar disorder. *Neuropsychopharmacology* 33: 110-133, 2008.

Sherman H. *Your Power to Heal*. New York: Harper and row, 1972.

Sheldon W.H. The varieties of temperament: a psychology of constitutional differences. NewYork: Harper and Brothers, 1942.

Shoemaker S. Realization and Mental Causation, in Gillett and Loewer 2001, pp. 74–98. Reprinted in Shoemaker 2003, pp. 427–51.

Sinha R., Lovallo W.R. and Parsons O.A. Cardiovascular Differentiation of emotions. *Psychosomatic Medicine* 54, 1992.

Skinner B.F. *Contingencies of reinforcement: a theoretical analysis.* New York: Meredith Corporation, 1969. French trans. *L'analyse expérimentale du comportement.* Bruxelles: Mardaga, 1971.

Smart J.J.C. Sensations and Brain Processes, *Philosophical Review*, 68: 141–156, 1959.

Soon et al. *Unconscious determinants of free decisions in the human brain, Nature Neuroscience 11, p.543 – 545,* 2008.

Sperry R. *Science and Moral Priority*, Columbia University Press, 1983.

Spinoza B. *Ethics III* 1677. French trans. Ch. Appuhn: *Œuvres III, Ethique, démontrée suivant l'ordre géométrique et divisée en cinq parties.* Paris: Garnier-Flammarion, 1965.

Strevens, M. Comments on Woodward, Making Things Happen, *Philosophy and Phenomenological Research* 77 (1):171-192, 2008.

Sudarshan E.C.G. Perception in Quantum Systems, in A. van der Merwe ed. *Old and new questions in physics, cosmology, philosophy and theoretical biology*, p. 457-467. New York Plenum, 1983.

Szekely E.B. The Discovery of the Essene Gospel of Peace. The Essenes and the Vatican. Academy Books. San Diego, California, 1975.

Takaesu Y. Circadian rhythm in bipolar disorder: A review of the literature. Psychiatry and Clinician Neurosciences 72: 673-682, 2018.

Taylor G.J., Bagby R.M. and Parker J.D.A. *Disorders of affect regulation. Alexithymia in medical and psychiatric illness.* Cambridge: Cambridge University Press, 1997.

Teilhard de Chardin P. *Le phénomène humain* (1955). Re-published by Le Seuil, Col. Point Essais, 1978.

Temoshok L. and Hefler B.M. Introducing the 'Type B' constellation into psychological ontology: Theory, measurement, and evidence for validity. Washington D.C., American Psychological Association, 1983.

Tittel W., Brendel J., Gisin B., Herzog T.; Zbinden H. and Gisin N. (1998) Experimental demonstration of quantum-correlations over more than 10 kilometers, *Physical Review A, 57.*

Tononi G. The Information Integration Theory of Consciousness, in *The Blackwell Companion To Consciousness.* Edited by Max Velmans and Susan Schneider, 2007.

Tressoldi P.E., Massacceci S., Martinelli M. and Cappato S. Mental Connection at Distance: Useful for Solving Difficult Tasks? *Psychology* Nov. 2011.

Tressoldi P.E., Khrennikov A. Remote state preparation of mental information: a theoretical model and a summary of experimental evidence. *NeuroQuantology* 10 (3): 394-402, 2012.

Trombini G. and Baldoni F. *La psychosomatique*. Coll. Psycho. Paris: Inpress, 2005.

Tsirelson B.S. Quantum generalizations of Bell's inequality. Letters in Mathematical Physics 4, 93(100), 1980.

Tung A. and Ric M.P. How Antidepressant and Antipsychotic Medications Work. *"Medications" issue of Visions Journal*, 2007, 4 (2), pp. 7-8, 2007.

Uehara M. and Koibuchi S.K. Thermodynamics of the Heart, in Application of Thermodynamics to Biological and Materials Science, ed. M. Tadashi, InTech 2011.

Uzan, P. *Conscience et physique quantique.* Paris: Vrin, 2013.

Uzan P. On the Nature of Psychophysical Correlations *Mind and Matter* 12 (1), 2014a.

Uzan P. The Quantum-like Approach to Psychosomatic Phenomena in Application, *Axiomathes* 25 (3), 2014b.

Uzan P. Psychologie cognitive et calcul quantique *Implications Philosophiques*, 2014c.

Uzan P. Complementarity in Psychophysics, in *Lecture Notes in Computer Science,* Ed. Filk and Atmanspacher, Springer, Vol. 9535: 168-178, 2016.

Uzan P. Deciding the Mind-Body Problem Experimentally, *Axiomathes*, 27: 333-354, 2017.

Uzan P. Super-Quantum Correlations: A Necessary Clarification. *Journal of Quantum Information Science,* Vol.8 No.3, 2018.

Uzan P. Why Super-Quantum, Non-Signaling Correlations Cannot Exist", *Physics Essays* Vol. 33 No 2, 2020.

Vasant L. Textbook of Ayurveda: Fundamental Principles. Albuquerque (Vol. 1), New Mexico: Ayurvedic Press, 2002.

Vasiliev L. Mysterious Phenomena of the Human Psyche. New Hyde Park, NY: University Bookes, 1965.

von Franz M.L. *On Divination and Synchronicity, the Psychology of Meaningful Chance.* Inner City Books, 1980.

von Franz M.L. *Number and Time*. Northwestern University Press, 1974.

Walach H., Tressoldi P. and Pederzoli L Mental, Behavioural and Physiological Nonlocal Correlations within the Generalized Quantum Theory Framework: A Review. *SSRN,* 2016.

Walker, E. H. Foundations of Paraphysical and Parapsychological Phenomena, in L. Oteri (ed.) *Quantum Physics and Parapsychology*, New York: Parapsychology Foundation, 1975.

Watson, J.B. *Behaviourism.* New York, NY: W.W. Norton & Company, 1925.

Watzlawick P., Beavin J.H. and Jackson D. Une logique de la communication. Paris: Seuil, 1967.

Weber E. H. *De subtilitate tactus*. In *De pulsu, resorptione, auditu et tactu* (pp. 44–174) Leipzig: Koehler, 1834.

Wegner, D.M. and Wheatley, T. Apparent mental causation: Sources of the experience of will. *American Psychologist*. 54 (7): 480–492, 1999.

Weiner H, Thaler M, Reiser M.F. and Mirsky T.A. Etiology of duodenal ulcer: Relation of specific psychological characteristics to rate of gastric secretion (serum pepsinogen). *Psychosomatic Medicine* 19: 1-10, 1957.

Weiner N. Or control and communication in the animal and the machine. *Massachusetts Institute of technology press,* 1962.

Weyl H. *Philosophy of Mathematics and Natural Science.* Princeton University Press, Princeton, 1949.

Whitehead A.N. *Process and Reality: An Essay in Cosmology* (1929). Corrected and published by Griffin D.R. and Sherburne D.W., Free Press, 1978.

Wigner E.P. Remarks on the Mind-Body question, in *Quantum Theory of Measurement*, J. Wheeler and W. Zurek, eds., Coll. Princeton Legacy Library, Princeton University Press, 1983.

Williams P. *Rethinking Madness*. Sky's Edge Publishing, 2012.

Wilson W.P. Mental health benefits of religious salvation. Dis. Nerv. Syst. 33: 382-386, 1972.

Winawer J. A motion aftereffect from visual imagery of motion. *Cognition* 114:276–284, 2010.

Woodward, J. *Making Things Happen*, Oxford University Press, 2003.

Woodward, J. Interventionism and Causal Exclusion, http://philsciarchive pitt.edu/8651, 2011.

Wu W.H., Bandilla E., Ciccone D.S. et al. Effects of qigong on late-stage complex regional pain syndrome. *Altern Ther Health Med* (1):45-54, 1999.

Wu JC, Kelsoe JR, Schachat C, et al. Rapid and sustained antidepressant response with sleep deprivation and chronotherapy in bipolar disorder. *Biol Psychiatry* 66:298-301, 2009.

Zahbib K, Milhiet V, Jamain S, Bellivier F. Santé physique et trouble bipolaire. Annales Médico-Psychologiques 170:56–61, 2012.

Zurek W.H. Pointer basis of quantum apparatus: Into what mixture does the wave packet collapse? Phys. Rev. D24, 1981.

Zurek W.H. Environment-induced superselection rules. *Physical Review* D26, 1982.

www.ingramcontent.com/pod-product-compliance
Ingram Content Group UK Ltd.
Pitfield, Milton Keynes, MK11 3LW, UK
UKHW020240040925
462575UK00004B/167
* 9 7 8 3 0 3 0 9 0 1 7 2 1 *